Lecture Notes in Mathematics

Edited by A. Dold and B. Eckmann

1026

Wilhelm Plesken

Group Rings of Finite Groups Over p-adic Integers

Springer-Verlag
Berlin Heidelberg New York Tokyo 1983

Author

Wilhelm Plesken
Lehrstuhl D für Mathematik, RWTH Aachen
Templergraben 64, 5100 Aachen, Federal Republic of Germany

AMS Subject Classifications (1980): 16A18, 16A26, 16A64, 20C05, 20C11, 20C20

ISBN 3-540-12728-3 Springer-Verlag Berlin Heidelberg New York Tokyo
ISBN 0-387-12728-3 Springer-Verlag New York Heidelberg Berlin Tokyo

Library of Congress Cataloging in Publication Data. Plesken, Wilhelm, 1950- Group rings of finite groups over p-adic integers. (Lecture notes in mathematics; 1026) Bibliography: p. Includes index. 1. Group rings. 2. Finite groups. 3. p-adic numbers. I. Title. II. Series: Lecture notes in mathematics. (Springer-Verlag); 1026. QA3.L28 no. 1026 [QA171] 510s [512'.22] 83-16985
ISBN 0-387-12728-3 (U.S.)

Printing and binding: Beltz Offsetdruck, Hemsbach/Bergstr.
2146/3140-543210

PREFACE

In the present notes the theory of orders over Dedekind domains is applied to study group rings of finite groups over the p-adic integers. The presentation grew out of my Habilitationsschrift at the Rheinisch-Westfälische Technische Hochschule Aachen, but goes far beyond it. The major part of the material is accessible to anyone who knows the definition of a maximal order and is familiar with the elements of modular representation theory of finite groups.

It was Professor H. Zassenhaus who introduced me to the subject a couple of years ago, and it is fair to say that these notes would have never been written without him. I acknowledge with pleasure that I greatly profited from discussions with H. Benz, H. Jacobinski, G. Michler, H. Pahlings, and K. Roggenkamp. I am very grateful to W. Rump for reading the manuskript and to a referee for pointing out an error in an earlier version, the correction of which lead to generalizations of some results. For typing the manuscript I would like to thank Mrs. D. Burkel, Miss A. Nijenhuis, and Mrs. C. Schneider. Part of the work was done while I held a Heisenberg scholarship. I would like to thank the Deutsche Forschungsgemeinschaft for the opportunities given to me by this grant. Finally I would like to thank the Sonderforschungsbereich für theoretische Mathematik, Bonn, for their hospitality during part of the preparation of the manuscript.

W. Plesken

CONTENTS

I. Introduction

The knowledge of the structure of the group ring RG of a finite group G over the ring R of integers in a finite extension field K of the p-adic number field $\mathbb{Q}_p$ yields insight into the possible actions of G on abelian groups. Compared with the matrix ring $\Omega^{n\times n}$ over the maximal R-order Ω in a K-division algebra D the group ring RG is little understood in case p divides the order $|G|$ of the group G . The present paper makes one step towards a description of RG in terms of such matrix rings, their twosided ideals, and isomorphisms between certain (finite) factor rings. Such a description more or less allows to read off the isomorphism types of the irreducible RG-lattices and their possible embeddings into one another, as well as a way the projective indecomposable RG-lattices are built up form certain irreducible ones. (An RG-lattice L is called irreducible if $KL := K\otimes_R L$ is an irreducible KG-module.) Among other examples, the principal 2-block of the group ring of $SL_2(q)$, q an odd prime power, over the 2-adic integers, and blocks with cyclic defect groups over the ring of p-adic integers are treated as applications of the general theory.

As BRAUER [Bra 56] points out, the most interesting arithmetic properties of $\Lambda = RG$ are lost when one passes from RG to a maximal R-order in $A = KG$ which contains RG . Since hereditary orders are equally well understood as maximal orders, JACOBINSKI, cf. [Jac 66], [Jac 81], suggests to embed Λ into a certain hereditary R-order in A , which he calls a hereditary hull of Λ . In Chapter II of this paper the hereditary orders are replaced by the considerably more general graduated orders or even graduable orders, cf. ZASSENHAUS [Zas 75], as the "well-known" R-orders in which Λ might be embedded and a graduated hull of Λ is defined. Of course Λ need not be a group ring for this but only an arbitrary R-order in a semisimple K-algebra A. Graduated orders in A are essentially defined by the property that they contain a full set of primitive orthogonal idempotents of A (cf. (II.1) for the exact definition) and are distinguished by the property that they are determined by their irreducible lattices, cf. (II.8).

There are several advantages of the replacement of hereditary hulls by graduated hulls. Graduated orders can still be described by comparatively few invariants, cf. (II.2) and (II.3). A graduated hull is generally a better approximation (from above) to the order Λ than a hereditary hull. Indeed, it follows from (II.8) that there exists only one unique graduated hull under certain conditions in which case it is the intersection of all hereditary hulls or equivalently of all maximal R-orders containing Λ. Thirdly JACOBINSKI's conductor formula for a hereditary R-order containing a group ring $\Lambda = RG$, cf. [Jac 66], [Jac 81], can be generalized in two respects: The hereditary order can be replaced by a graduated order containing Λ and Λ need not be a group ring but only a "selfdual order" with respect to a generalized trace bilinear form on the separable K-algebra A, (cf. (III.1) for a proper definition). The conductor of some overorder Γ of a selfdual order Λ, i.e. the biggest Γ-ideal contained in Λ, turns out to be the dual Γ^* of Γ with respect to the generalized trace bilinear form of A belonging to Λ, cf. (III.7). The conductor formula (III.8) gives an explicit description of the conductor of Γ in Λ in terms of the structural invariants of Γ in case Γ is a graduated order. The idea is that a graduated hull Γ of Λ restricts the possibilities for Λ considerably more than a hereditary hull, since $\Gamma^* \subseteq \Lambda \subseteq \Gamma$, and that the conductor formula makes it possible to discuss these inclusions. Note, if $\Gamma = \bigoplus_{s=1}^{h} \varepsilon_s\Lambda$, where $\varepsilon_1,\ldots,\varepsilon_h$ are the central primitive idempotents of A, then the conductor is also equal to $\bigoplus_{s=1}^{h} (\varepsilon_s\Lambda\cap\Lambda)$. Other examples of selfdual orders apart from group rings are twisted group rings, cf. (III.2), (III.4), and orders of the form $\varepsilon\Lambda\varepsilon$ where ε is an idempotent in a selfdual R-order Λ.

These ideas can most successfully be applied to group rings RG where R is a sufficiently large unramified (finite) extension of the p-adic integers $\mathbb{Z}_p$. But of course the primary group theoretical interest is concentrated in the group rings $\mathbb{Z}_pG$ over the p-adic integers, for instance because they determine the possible actions of G on finite abelian p-groups. To develop the tools for the Galois descent from RG to $\mathbb{Z}_pG$ Chapter II.b discusses graduable orders, i.e. orders which become graduated orders after a sufficiently big unramified ground ring extension, cf. Definition (II.13). Among other characterizations Theorem (II.16), cf. also (III.12), gives an easily applicable criterion for $\varepsilon_s\Lambda$ to be a graduable order: Certain modified (cf.(III.10)) decomposition numbers have to be equal to zero or one. As for the Galois descent Theorem (II.20) gives a satisfactory answer: A graduable order Γ in a central simple algebra is determined up to isomorphism by $\Gamma' = R'\otimes_R\Gamma$ and the natural embedding $\Gamma/\mathrm{Jac}(\Gamma)\hookrightarrow\Gamma'/\mathrm{Jac}(\Gamma')$ for some arbitrary unramified extension R' of R. Many things done with graduated orders in later chapters, could also be done with graduable orders, e.g. the conductor formula could be proved for graduable overorders. Since this can easily be obtained from the stated results and (II.20) it is usually not mentioned explicitely. This is about as far as the general theory is developed in Chapter II and III, the main results being the characterizations of graduated and graduable orders in (II.8) and (II.16), the essential uniqueness for the Galois descent for graduated to graduable orders in (II.20) and of course the conductor formula (III.8). It should be noted, however, that Chapter II.b can be skipped upon first reading, since the other chapters are kept essentially independent of this part.

Chapter IV discusses selfdual orders Λ for which $\bigoplus_{s=1}^{h} \varepsilon_s\Lambda$ (ε_s as above) is a graduated R-order in A . This is essentially tantamount to demanding that the decomposition numbers of Λ are all equal to 0 and 1 , (and R "big enough") cf. (III.12) or Chapter IIb for the exact conditions. In this situation, the projective indecomposable Λ-lattices P_i have the property that $\varepsilon_s P_i$ is irreducible or 0 by Brauer's reciprocity. The major part of Chapter IV is a careful analysis of the embedding of P_i in the completely decomposable Λ-lattice $\bigoplus_{s=1}^{h} \varepsilon_s P_i$ by investigating the "amalgamating factors" $(\bigoplus_{s=1}^{h} \varepsilon_s P_i)/P_i$ and $P_i/\bigoplus_{s=1}^{h} (\varepsilon_s P_i \cap P_i)$ of P_i . For most applications in later chapters (IV.1), (IV.7), and (IV.10) to (IV.12) are sufficient.

The last four chapters contain applications of the theory developed in the first chapters to the explicite computation of group rings in the sense described at the beginning of this introduction. The general procedure consists of two steps: At first one determines $\bigoplus_{s=1}^{h} \varepsilon_s\Lambda$ which essentially amounts to finding the sublattices of all irreducible Λ-lattices. Then one has to find the embedding of Λ into $\bigoplus_{s=1}^{h} \varepsilon_s\Lambda$, i.e. to see how the $\varepsilon_s\Lambda$ (resp. $\varepsilon_s P_i$) are amalgamted to Λ (resp. P_i). In practice these two steps are not performed one after the other. Indeed, it is part of the idea to use step 2 to compare the various epimorphic images $\varepsilon_s\Lambda$ of Λ and thereby getting information to perform step 1. Of course, step 2 can only be completed after step 1 is fully carried out.

Chapter V discusses block ideals of group rings with all diagonal Cartan numbers equal to 2. In case all Frobenius characters are real a complete description of the ringtheoretical structure of these blocks is given. In Chapter VI step 1 of the procedure outlined above is carried out for the principal blocks of various group rings with decomposition numbers 0 and 1 (i.e. the sublattices of the irreducible lattices in the block are determined). The examples are the symmetric group S_{10} at the prime 5, $SL_3(3)$ at the prime 2, $SL_3(4)$ at the prime 3, and the Mathieu group M_{11} at the primes 2 and 3. The usual information one starts out with in Chapter VI and VII is the character table of the group and the decomposition numbers. Of course, some insight in the subgroup structure is always a help. In Chapter VII the principal blocks of RG, $G = SL_2(q)$ with q an odd prime power and R the ring of 2-adic integers are completely determined in the above sense. For instance for $q \equiv \pm 1 \pmod 8$, it turns out that all irreducible lattices of this block are uniserial except possibly for the ones of R-rank q belonging to the Steinberg character.

Finally blocks with cyclic defect groups are described in Chapter VIII, where the Dedekind ring R is first assumed to be a sufficiently large unramified extension of the p-adic integers. In particular a generalization of Brauer's Theorem 11 in [Bra 41] on the sublattices of the irreducible lattices in a block of defect 1 is proved, cf. (VIII.3). This was mentioned as an open problem at the end of Dade's basic paper [Dad 66] on blocks with cyclic defect groups, cf. also [Fei 82]. As a corollary one obtains linear congruences for the central characters modulo $|D|R$ where D is a cyclic defect group of the block. In a final step the description is extended to blocks with cyclic defect groups over the ring of p-adic integers.

Some comments on the earlier literature might be useful. Graduated orders turn up at various places under various names; e.g. in [Jat 74] their global dimensions are investigated, cf. also [WiR 82]; in [ZaK 77] the graduated orders of finite representation type were characterized, cf. also [Rum 81a], [Rum 81b]; the connection of graduated orders with orders Λ whose irreducible lattices have a distributive lattice of Λ-sublattices was observed in [Ple 77], cf. also [Zas 69], [Ple 80a], [Rum 81b]. The major part of the theory developed here is already contained in the author's Habilitationsschrift [Ple 80a] for group rings RG under the two additional assumptions that the quotient field K of R is a splitting field for the group G and that the decomposition numbers are all equal to 0 and 1; cf. also [Ple 80b] for a short summary of those results. The expression for $\bigoplus_{s=1}^{h} (\varepsilon_s RG \cap RG)$ there was obtained as a consequence of Schur's relations, cf. [Hup 67] page 477, which allow to express matrix units in the group algebra KG as linear combinations of the group elements with coefficients coming from irreducible matrix representations of G, cf. also [Ser 77]. The presentation given here was influenced (at a late stage) by Jacobinski's lectures on hereditary orders, cf. [Jac 81] and his discussion of blocks of defect 1. For other approaches to special blocks of group rings cf. also [Rog 80a], [Rog 80b], [Rog 81].

The used notation is standard. Groups usually act from the left side, module homomorphisms are written on the left, right resp. left side for right, left, resp. bi-modules. All rings have a unit and all modules are unital. If not stated otherwise, all modules are left finitely generated modules (and hence also viewed as right modules over their endomorphism rings). A general reference for orders over Dedekind domains is [Rei 75].

II. Graduated and graduable orders

In this chapter R is a complete local Dedekind domain with quotient field K and maximal ideal p. Furthermore D is a (finite dimensional) separable division algebra over K, Ω the maximal R-order of D, $\mathfrak{P} = \mathrm{Jac}(\Omega)$ the maximal ideal of Ω (cf. e.g. [Jac 81], [Rei 75]). A will denote a separable K-algebra containing an R-order Λ such that $K\Lambda = A$.

II.a. Definition and characterization of graduated orders

(II.1) Definition. *Λ is called a graduated order if there exist orthogonal (primitive) idempotents $\varepsilon_1, \ldots, \varepsilon_t$ in Λ with $1 = \varepsilon_1 + \ldots + \varepsilon_t$ $(1 = 1_\Lambda)$ such that $\varepsilon_i \Lambda \varepsilon_i$ is a maximal order in $\varepsilon_i A \varepsilon_i$ for $i = 1, \ldots, t$.*

In particular, the maximal orders and the hereditary orders are among the graduated orders. Note, if K is a splitting field of A, then Λ is a graduated order iff Λ contains a complete set of orthogonal idempotents of A. If A decomposes into a direct sum of minimal twosided ideals $A_1, \ldots, A_h$, then Λ is a graduated order if and only if Λ is the direct sum of the $\Lambda_s = \Lambda \cap A_s$ and each of the Λ_s is a graduated order in A_s $(s = 1, \ldots, h)$. Therefore it suffices to investigate graduated orders in simple algebras. Let A be the matrix ring $D^{n \times n}$ of degree n over D in the sequel. To describe the obvious examples of graduated orders, the following notation is useful: For $\tilde{n} = (n_1, \ldots, n_t) \in \mathbb{N}^{1 \times t}$ with $n = \sum_{i=1}^{t} n_i$ and $M = (m_{ij}) \in \mathbb{Z}^{t \times t}$ let

$\Lambda(\Omega,\tilde{n},M) = \{(a_{ij}) \mid a_{ij} \in (\mathfrak{p}^{m_{ij}})^{n_i \times n_j},\ 1 \leq i,j \leq t\} \subseteq D^{n\times n} = A.$

I.e., the elements of $\Lambda(\Omega,\tilde{n},M)$ are $n\times n$-matrices over D partitioned into $n_i \times n_j$-submatrices a_{ij} the entries of which lie in $\mathfrak{p}^{m_{ij}}$. If $\Omega = R$, one writes simply $\Lambda(\tilde{n},M)$ instead of $\Lambda(R,\tilde{n},M)$. Obviously $\Lambda = \Lambda(\Omega,\tilde{n},M)$ is an order in A, iff $M = (m_{ij})$ satisfies $m_{ii} = 0$ and $m_{ij} + m_{jk} \geq m_{ik}$ for $1 \leq i,j,k \leq t$. In this case Λ is already a graduated order, the matrix graduation being induced by the standard diagonal idempotents.

(II.2) Definition. A graduated order Λ in $A = D^{n\times n}$ is said to be in standard form, if there exist $\tilde{n} = (n_1,\ldots,n_t) \in \mathbb{N}^{1\times t}$ with $n_1 + \ldots + n_t = n$, $M = (m_{ij}) \in \mathbb{Z}_{\geq 0}^{t\times t}$ such that $\Lambda = \Lambda(\Omega,\tilde{n},M)$ and

$$(*)\quad \begin{cases} m_{ij} + m_{jk} \geq m_{ik} \\ m_{ii} = 0 \\ m_{ij} + m_{ji} > 0 \quad (i \neq j) \end{cases}$$

for $1 \leq i,j,k \leq t$. In this case M is called the exponent matrix of Λ and $\tilde{n}$ the dimension type of Λ.

Since any two complete sets of primitive idempotents of A can be conjugated by nonsingular matrices of A into each other and since the ideals of Ω are the powers of $\mathfrak{p}$, one easily verifies (II.3), cf. e.g. [Zas 75].

(II.3) Remark. Each graduated order Λ in $A = D^{n\times n}$ is isomorphic to a graduated order in standard form.

For a graduated order in standard form it is easy to compute the irreducible lattices. Therefore the proof of (II.4) is left to the reader, cf. e.g. [Zas 75], [Ple 77], [Ple 80a], [Rum 81b].

<u>*(II.4) Remark.*</u> *Let* $\Lambda = \Lambda(\Omega,\tilde{n},M)$ *be a graduated order of standard type in* A .

(i) The Jacobson radical of Λ *is given by* $Jac(\Lambda) = \Lambda(\Omega,\tilde{n},M+I_t)$, *where* I_t *is the* $t\times t$*-unit matrix.*

(ii) $\Lambda/Jac(\Lambda) \cong \bigoplus_{i=1}^{t} (\Omega/\mathfrak{P})^{n_i\times n_i}$.

(iii) Let $V = D^{n\times 1}$ *be the standard irreducible module of* $A = D^{n\times n}$. *The set* $\mathfrak{Z}(V)$ *of all* Λ*-lattices* $\neq 0$ *in* V *is given by all*

$$L(\tilde{m}) = \left\{ \begin{pmatrix} a_1 \\ \vdots \\ a_t \end{pmatrix} \mid a_i \in (\mathfrak{P}^{m_i})^{n_i\times 1} \right\} \subseteq V$$

with $\tilde{m} = (m_1,\ldots,m_t)^{tr} \in \mathbb{Z}^{n\times 1}$ *satisfying*

(**) $m_{ij} + m_j \geq m_i \quad (1 \leq i,j \leq t)$

(iv) Two Λ*-lattices* $L(\tilde{m}_1)$, $L(\tilde{m}_2) \in \mathfrak{Z}(V)$ *are isomorphic if and only if* $\tilde{m}_1 - \tilde{m}_2$ *is a multiple of* $(1,\ldots,1)^{tr} \in \mathbb{Z}^{t\times 1}$, *i.e.* $L(\tilde{m}_1) = L(\tilde{m}_2)\mathfrak{P}^{\alpha}$ *for some* $\alpha \in \mathbb{Z}$.

(v) Each projective indecomposable Λ*-lattice is irreducible and isomorphic to* $L(M_i)$ *for some* $i = 1,\ldots,t$, *where* M_i *is the i-th column of* M .

(vi) Define $S_i = L(M_i)/Jac(\Lambda)L(M_i)$ *for* $i = 1,\ldots,t$. *Then* $S_1,\ldots,S_t$ *form a set of representatives of the simple* Λ*-(torsion)modules. For solutions* $\tilde{m}_1,\tilde{m}_2$ *of* (**) *with* $L(\tilde{m}_1) \subseteq L(\tilde{m}_2)$ *the i-th coefficient of* $\tilde{m}_1 - \tilde{m}_2$ *is the multiplicity of* S_i *in a composition series of the* Λ*-module* $L(\tilde{m}_2)/L(\tilde{m}_1)$ *for* $i = 1,\ldots,t$.

(vii) Each injective indecomposable Λ*-lattice is irreducible and isomorphic to* $L({}_iM)$ *for some* $i = 1,\ldots,t$, *where* ${}_iM$ *is the i-th column of* $-M^{tr}$. (A Λ-lattice L is injective, if $\mathrm{Hom}_R(L,R)$ is a projective right Λ-lattice.)

(viii) The two-sided (fractional) ideals of Λ *in* A *are given by* $\Lambda(\Omega,\tilde{n},N)$, *where* $N = (n_{ij}) \in \mathbb{Z}^{t\times t}$ *satisfies* $m_{ij} + n_{jk} \geq n_{ik}$ *and* $n_{ij} + m_{jk} \geq n_{ik}$ *for* $1 \leq i,j,k \leq t$.

Note, by part (ii) of this remark the dimension type $\tilde{n}$ depends only on the isomorphism type of Λ (up to the order of the n_i). By (II.4) (iii) and (iv) the irreducible Λ-lattices of the graduated order $\Lambda = \Lambda(\Omega,\tilde{n},M)$ fall into finitely many isomorphism classes the number of which only depends on M . A convenient set of representatives can be produced as follows: For $\tilde{m} = (m_1,\ldots,m_t)^{tr} \in \mathbb{Z}^{t\times 1}$ satisfying (**) let

$$\mathfrak{R}(\tilde{m}) = \{L(\tilde{l}) \mid \tilde{l} = (l_1,\ldots,l_t)^{tr} \in \mathbb{Z}^{t\times 1} \text{ satisfying (**)},$$
$$l_i \geq m_i \text{ for } i = 1,\ldots,t\,;\ l_j = m_j \text{ for one } j = 1,\ldots,t\} =$$
$$= \{L \in \mathfrak{J}(V) \mid L \subseteq L(\tilde{m}) \,,\ L \not\subseteq L(\tilde{m})\mathfrak{P}\}\ .$$

This set $\mathfrak{R}(\tilde{m})$ can conveniently be computed "layer by layer" as follows:
Compute the (at most t) maximal Λ-sublattices of $L(\tilde{m})$ and drop those which are not in $\mathfrak{R}(\tilde{m})$. Continue with the maximal Λ-sublattices of $L(\tilde{m})$ in $\mathfrak{R}(\tilde{m})$ which are contained in $\mathfrak{R}(\tilde{m})$ to get the second maximal Λ-sublattices of $L(\tilde{m})$ in $\mathfrak{R}(\tilde{m})$ etc.. (Note, $L(\tilde{m})$ contains a maximal sublattice L with $L(\tilde{m})/L \cong A_i$ iff $m_{ij} + m_j > m_i$ for $j = 1,\ldots,t$, $j \neq i$.)

Unlike the dimension type $\tilde{n}$ the exponent matrix M of a graduated order Λ is not uniquely defined by the isomorphism type of Λ , because Λ can well be isomorphic to more than one graduated order in standard form. Therefore ZASSENHAUS [Zas 75] introduced structural invariants m_{ijk} of a graduated order Λ in $A = D^{n\times n}$ as follows: Lift the central primitive idempotents $\bar{\varepsilon}_1,\ldots,\bar{\varepsilon}_t$ of $\Lambda/\mathrm{Jac}(\Lambda)$ to orthogonal idempotents $\varepsilon_1,\ldots,\varepsilon_t$ of Λ with $1 = \varepsilon_1 + \ldots + \varepsilon_t$. Let $\Lambda_{ij} = \varepsilon_i \Lambda \varepsilon_j$ for $i,j = 1,\ldots,t$. Then there are nonnegative numbers $m_{ijk} \in \mathbb{Z}$ satisfying

$$\Lambda_{ij}\Lambda_{jk} = \mathfrak{p}^{m_{ijk}}\Lambda_{ik} \quad \text{for} \quad i,j,k = 1,\dots,t \ .$$

(Note Λ_{ii} is a maximal order for $i = 1,\dots,t$.) The m_{ijk} satisfy

$$(***) \quad \begin{cases} m_{ijk} + m_{ikl} = m_{ijl} + m_{jkl} \\ m_{iii} = 0 \ (= m_{iij} = m_{ijj}) \\ m_{iji} > 0 \quad (j \neq i) \end{cases}$$

for $1 \leq i,j,k,l \leq t$. The first equations of (***) follow from associativity: $(\Lambda_{ij}\Lambda_{jk})\Lambda_{kl} = \Lambda_{ij}(\Lambda_{jk}\Lambda_{kl})$, the second from the property of Λ_{ii} to be an order, and the third from the choice of the idempotents ε_i . If $\varepsilon_1',\dots,\varepsilon_t'$ is a second set of idempotents of Λ such that the $\varepsilon_i' + \mathrm{Jac}(\Lambda)$ are the central primitive idempotents of $\Lambda/\mathrm{Jac}(\Lambda)$ the structural invariants m_{ijk}' of Λ with respect to the ε_i' are obtained by permuting the indices of the m_{ijk} . Namely there exists an inner automorphism α of Λ and a permutation $\pi \in S_n$ with $\alpha(\varepsilon_i) = \varepsilon_{\pi i}'$ for $i = 1,\dots,t$. Therefore $m_{ijk} = m_{\pi i,\pi j,\pi k}'$, $1 \leq i,j,k \leq t$. Hence the structural invariants (up to order) depend only on the isomorphism type of the graduated order Λ . Call $\tilde{n} = (n_1,\dots,n_t)$ the dimension type of Λ , where n_i is the unique natural number with $\Lambda_i \cong \Omega^{n_i \times n_i}$, $i = 1,\dots,t$.

If $\Lambda = \Lambda(\Omega,\tilde{n},M)$ is a graduated order in standard form, then clearly $\tilde{n}$ is also the dimension type of Λ in the sense just defined and the structural invariants of Λ are given by

$$(****) \quad m_{ijk} = m_{ij} + m_{jk} - m_{ik} \qquad \text{for} \quad 1 \leq i,j,k \leq t \ .$$

(II.5) Lemma. Let $M' = (m_{ij}') \in \mathbb{Z}_{\geq 0}^{t\times t}$ be a second solution of (****). Then there are integers $m_1,\dots,m_t \in \mathbb{Z}$ with $m_{ij}' = m_{ij} + m_i - m_j$ for $1 \leq i,j \leq t$.

Proof: Let $x_{ij} = m_{ij}' - m_{ij}$ for $1 \leq i,j \leq t$. Then (****) implies

$x_{ij} + x_{jk} - x_{ik} = 0$ for $1 \leq i,j,k \leq t$. Summation over k yields $tx_{ij} = y_i - y_j$ for $1 \leq i,j \leq t$, where the y's are defined by $y_i = \sum_{k=1}^{t} x_{ik}$. The $m_i = [t^{-1}y_i]$ $(1 \leq i \leq t)$ satisfy the required condition, where $[x]$ denotes the biggest integer smaller or equal to x for $x \in \mathbf{R}$.

q.e.d.

By this lemma a graduated order $\Lambda' = \Lambda(\Omega,\tilde{n},M')$ in standard form with $(m'_{ij}) = M'$ satisfying (****) is isomorphic to $\Lambda = \Lambda(\Omega,\tilde{n},M)$, namely $\sum_{i=1}^{t} p^{m_i}\varepsilon_i$ transforms Λ into Λ', where $\varepsilon_1,\ldots,\varepsilon_t$ are the (standard diagonal) idempotents of Λ such that $\varepsilon_i + \mathrm{Jac}(\Lambda)$ are the central primitive idempotents of $\Lambda/\mathrm{Jac}(\Lambda)$, $m_i \in \mathbf{Z}$ are as in (II.5) and p is a generator of the maximal ideal $\mathfrak{P}$ of Ω. Hence, two graduated orders with the same dimension type and the same structural invariants are isomorphic.

__(II.6) Proposition__ (cf. [Zas 75]). (i) Two graduated orders Λ *and* Λ' *in* $A = D^{n\times n}$ *with dimension types* $(n_1,\ldots,n_t)$, $(n'_1,\ldots,n'_{t'})$ *and structural invariants* $(m_{ijk})_{1\leq i,j,k\leq t}$ *resp.* $(m'_{ijk})_{1\leq i,j,k\leq t'}$ *are isomorphic if and only if* $t = t'$, $n_i = n'_{\pi i}$ $(1 \leq i \leq t)$, *and* $m_{ijk} = m'_{\pi i,\pi j,\pi k}$ $(1 \leq i,j,k \leq t)$ *for a suitable permutation* $\pi \in S_t$. *They are Morita equivalent if and only if* $t = t'$ *and* $m_{ijk} = m'_{\pi i,\pi j,\pi k}$ $(1 \leq i,j,k \leq t)$ *for a suitable permutation* $\pi \in S_k$.

(ii) For any family $(m_{ijk})_{1\leq i,j,k\leq t}$ *of nonnegative integers satisfying (***) and any* $\tilde{n} = (n_1,\ldots,n_t) \in \mathbf{N}^{t\times 1}$ *there exists a graduated order* Λ *in* $A = D^{n\times n}$ *with* $n = \sum_{i=1}^{t} n_i$ *such that* $\tilde{n}$ *is the dimension type of* Λ *and* m_{ijk} *are the structural invariants. For instance,* $\Lambda = \Lambda(\Omega,\tilde{n},M)$ *with* $M = (m_{ij}) \in \mathbf{Z}_{\geq 0}^{t\times t}$ *defined by any of the following equations*

(I_{k_0}) $\quad m_{ij} = m_{k_0 ij} \quad (1 \leq i,j \leq t)$ for a fixed $k_0 \in \{1,\dots,t\}$ or

(P_{k_0}) $\quad m_{ij} = m_{ijk_0} \quad (1 \leq i,j \leq t)$ for a fixed $k_0 \in \{1,\dots,t\}$.

Proof: (i) The first statement follows from the preceeding remarks and (II.3). The statement about Morita equivalence is a consequence of this.

(ii) One only has to verify (*) and (****) for the m_{ij} defined in this way. These are consequences of $m_{ijk} \geq 0$ $(1 \leq i,j,k \leq t)$ and (***).

q.e.d.

(II.6) says that the Morita equivalence types of graduated orders Λ in matrix rings over D are parametrized by the S_t-orbits on the nonnegative solutions of (***), where t is the number of components of $\Lambda/\mathrm{Jac}(\Lambda)$.

Since graduated orders are determined by their irreducible lattices, the rôle of irreducible lattices can easily be derived from the preceding discussion:

For a graduated R-order Λ in $A = D^{n \times n}$ with dimension type $\tilde{n} = (n_1,\dots,n_t)$ and structural invariants $(m_{ijk})_{1 \leq i,j,k \leq t}$ one has bijections between the three sets containing

(i) the maximal R-orders in A containing Λ ;

(ii) the isomorphic classes of irreducible Λ-lattices; or

(iii) the solutions $(m_{ij}) \in \mathbb{Z}_{\geq 0}^{t \times t}$ of (****) respectively.

Namely if $\Lambda = \Lambda(\Omega,\tilde{n},(m'_{ij}))$ is given a standard form, then (i) contains all $\Lambda(\Omega,\tilde{n},(m_i - m_j)_{1 \leq i,j \leq t})$ with $m_1,\dots,m_t$ satisfying (**) (with m_{ij} replaced by m'_{ij}); the isomorphism classes of (ii) are represented by $\mathfrak{R}(\tilde{m}_0)$ for some fixed $\tilde{m}_0$ satisfying (**) (cf.

discussion following (II.4)) and (iii) is given by $m_{ij} = m'_{ij} + m_i - m_j$ with $m_1, \ldots, m_t$ satisfying (**). In the language of these bijections the exponent matrices given in (II.6)(ii) correspond to the injective indecomposable Λ-lattices in the first case (I_{k_0}), and to the projective indecomposables in the second case (P_{k_0}) (which are all irreducible by (II.4)).

A useful algebraic interpretation of the m_{ij} and the m_{ijk} follows from (II.4)(vi): Let $S_1, \ldots, S_t$ be representatives of the simple Λ-(torsion)modules (in the natural order, i.e. $\Lambda_{ii} S_i = S_i$.). If L is an irreducible Λ-lattice which gives rise to the exponent matrix (m_{ij}) as discussed above, then m_{ij} is the multiplicity of S_i in a composition series of the Λ-torsion module L/P_j, where P_j is the unique projective cover of S_j in the set $\mathfrak{R}(L) = \{X \mid X \leq L \ , \ X \not\leq L\mathfrak{p}\}$ of representatives of irreducible Λ-lattices. Since the m_{ijk} for fixed k can be chosen to form the exponent matrix corresponding to the projective cover of S_k, one can use the same interpretation for the m_{ijk}.

A similar interpretation of the m_{ij} in terms of multiplicities of Λ-Λ-bimodules is possible: Let S_{ij} the simple Λ-Λ-module with $\Lambda_{ii} S_{ij} \Lambda_{jj} = S_{ij}$. Then m_{ij} is the multiplicity of S_{ij} in the Λ-Λ-torsion module $\mathrm{End}_{\Omega}(L)/\Lambda$, where L is an irreducible Λ-lattice corresponding to the exponent matrix (m_{ij}).

As one might have suspected, there is a more general formalism behind (***), (****), (II.5), and the bijections just discussed. This is the exactness of the following sequence of $\mathbb{Z} S_t$-lattices:

Let $\mathbb{Z}^{t^r}$ be the free $\mathbb{Z}$-module of all mappings of the r-fold cartesian product of $\{1, \ldots, t\}$ with itself into $\mathbb{Z}$ $(r \in \mathbb{Z}_{\geq 0})$.

There is an obvious action of the symmetric group S_t on $\mathbb{Z}^{t^r}$. The homomorphism $\delta_r : \mathbb{Z}^{t^r} \to \mathbb{Z}^{t^{r+1}}$ is defined to map $(m_{i_1,\ldots,i_r})_{1\le i_1,\ldots,i_r\le t}$ onto $(\bar{m}_{i_1\ldots i_{r+1}})_{1\le i_1,\ldots,i_{r+1}\le t}$, where $\bar{m}_{i_1\ldots i_{r+1}} = \sum_{k=1}^{r+1} (-1)^{r+1-k} m_{i_1,\ldots\hat{i}_k\ldots i_{r+1}}$ for $1\le i_1,\ldots,i_{r+1}\le t$. (The letter below ^ is to be omitted.) Clearly δ_r is a $\mathbb{Z}S_t$-homomorphism and $\delta_r\delta_{r+1}=0$. The exactness of the $\mathbb{Z}S_t$-sequence $0\to\mathbb{Z}^{t^0}\overset{\delta_0}{\to}\mathbb{Z}^{t^1}\overset{\delta_1}{\to}\mathbb{Z}^{t^2}\overset{\delta_2}{\to}\mathbb{Z}^{t^3}\to\ldots$ is a slight generalization of (II.6). For instance structural invariants are all elements $(m_{ijk})_{1\le i,j,k\le t}$ in the $\mathbb{Z}S_t$-lattice $\operatorname{im}\delta_2=\ker\delta_3$ of $\mathbb{Z}$-rank t^2-t+1 satisfying certain inequalities and $m_{iij}=m_{ijj}=0$. It would be interesting to have similar interpretations for $\operatorname{im}\delta_i=\ker\delta_{i+1}$ for $i>2$.

After the discussion of the parameters describing a graduated order Λ and its irreducible lattices now an analysis of those properties of graduated orders will be given, which distinguish them among general orders. It follows from (II.4)(ii) that the Λ-sublattices of an irreducible Λ-lattice satisfy the distributive laws with respect to taking intersections and sums. This motivates the subsequent discussion. Let V be a finite dimensional right D-vector space. A full right Ω-lattice in V is an Ω-lattice containing a D-basis of V. Call a set $\mathfrak{L}$ of full right Ω-lattices admissible, if the following three conditions are fulfilled:

(i) For $L_1,L_2\in\mathfrak{L}$ also $L_1\cap L_2\in\mathfrak{L}$ and $L_1+L_2\in\mathfrak{L}$;

(ii) for $L\in\mathfrak{L}$ and $\alpha\in\mathbb{Z}$, the lattice $L\mathfrak{p}^\alpha\in\mathfrak{L}$;

(iii) there is a constant $\delta\in\mathbb{N}$ such that for any two $L_1,L_2\in\mathfrak{L}$ there exists $\alpha\in\mathbb{Z}$ with $L_1\mathfrak{p}^\delta\subseteq L_2\mathfrak{p}^\alpha\subseteq L_1$.

Typical examples of admissible sets of Ω-lattices are the Λ-Ω-lattices (*0) in V, where Λ is an R-order in $\operatorname{End}_D(V)$, or the system

$\mathfrak{L}$ of Ω-lattices generated by a finite set of full Ω-lattices by taking sums, intersections, and multiplying with $\mathfrak{P}^{\alpha}(\alpha \in \mathbb{Z})$. Because of (i) admissible sets of Ω-lattices form a modular lattice with respect to intersection and sum, namely a sublattice of the lattice of all Ω-lattices in V. (iii) is a boundedness condition. An admissible system of Ω-lattices is said to have a system of compatible bases, if some $L \in \mathfrak{L}$ has an Ω-basis B such that one obtains an Ω-basis for each other $L' \in \mathfrak{L}$ by multiplying the basis vectors in B with elements of D, i.e. the transformation matrix can be chosen to be diagonal. The following proposition which connects the existence of a system of compatible bases with the distributivity of $\mathfrak{L}$ as a lattice, can be viewed as a generalization of (part of) the elementary divisor theorem (for local dedekind domains); namely one is in the situation of this classical result, if the admissable system is generated by two Ω-lattices in the sense explained above. In the case of commutative principal domains the result was proved in [Ple 77] and inspired by [Zas 69], where the connection of existence of simultanious decompositions of systems of subgroups of abelian groups and distributivity was noticed for the first time. In the case of vector space the connection had been observed in [Bre 75]. The proof of (II.7) given here differs from the ones in [Ple 77] and [Ple 80a]. (The proof in [Ple 80a] is in the spirit of the Hensel lemma starting out from the $\Omega/\mathfrak{P}$-vector space situation.) The result can also be obtained from [Rum 81a] and [Rum 81b].

(II.7) Proposition. Let $\mathfrak{L}$ be an admissible set of full (right) Ω-lattices in the (right) D-vector space V. $\mathfrak{L}$ has a system of compatible bases if and only if $\mathfrak{L}$ is distributive (as a lattice with respect to $\cap$ and $+$.).

Proof: If $\mathfrak{L}$ has a system of compatible bases, it is certainly

distributive (and due to the fact that the Ω-ideals in D form a distributive lattice, namely a chain). Conversely let $\mathfrak{L}$ be distributive. For $L \in \mathfrak{L}$ call $L' \in \mathfrak{L}$ $\mathfrak{L}$-maximal in L if $L' \subsetneqq L$ and for all $L'' \in \mathfrak{L}$ with $L' \subseteq L'' \subseteq L$ one has $L'' = L'$ or $L'' = L$. If L' is $\mathfrak{L}$-maximal in L then $L\mathfrak{P} \subseteq L' \subset L$. Since a distributive lattice cannot contain an intervall isomorphic to the subgroup lattice of the Klein four group, an induction shows $L/\bigcap_{i=1}^{k} L_i \cong \bigoplus_{i=1}^{k} L/L_i$ as $\Omega/\mathfrak{P}$-vector spaces, for any $L, L_1, \dots, L_k \in \mathfrak{L}$ with L_i $\mathfrak{L}$-maximal in L. In particular, any $L \in \mathfrak{L}$ contains only finitely many $\mathfrak{L}$-maximal sublattices. As a consequence of this and the boundedness condition (iii) for admissable sets, one obtains the finiteness of $\mathfrak{R}(L) = \{L' \in \mathfrak{L} \mid L' \subseteq L \,,\, L' \nsubseteq L\mathfrak{P}\}$ for all $L \in \mathfrak{L}$. Moreover, for any two $L_1, L_2 \in \mathfrak{L}$ one has a bijection between $\mathfrak{R}(L_1)$ and $\mathfrak{R}(L_2)$, which is induced by the multiplication of the Ω-lattices in $\mathfrak{R}(L_1)$ by suitable powers of $\mathfrak{P}$. (Note, $\mathfrak{R}(L) \times \mathbb{Z} \to \mathfrak{L} : (L', \alpha) \to L'\mathfrak{P}^{\alpha}$ is bijective.) For $L \in \mathfrak{L}$ denote by $\mathfrak{R}_1(L)$ the set of those Ω-lattices in $\mathfrak{R}(L)$ which have exactly one $\mathfrak{L}$-maximal sublattice. The above bijection between $\mathfrak{R}(L_1)$ and $\mathfrak{R}(L_2)$ $(L_1, L_2 \in \mathfrak{L})$ maps $\mathfrak{R}_1(L_1)$ onto $\mathfrak{R}_1(L_2)$. The core of the proof is the following:
Let $L = L_1, L_2, \dots, L_t, L_{t+1} = L\mathfrak{P} \in \mathfrak{L}$ such that L_{i+1} is $\mathfrak{L}$-maximal in L_i for $i = 1, \dots, t$. Then $\mathfrak{R}_1(L)$ consists precisely of $P_1, \dots, P_t$ where $P_i = \bigcap_{M \in \mathfrak{I}_i} M$ with $\mathfrak{I}_i = \{M \in \mathfrak{L} \mid M \subseteq L_i \,,\, M \nsubseteq L_{i+1}\}$ for $i = 1, \dots, t$.

To proof this, first observe $\mathfrak{I}_i \subseteq \mathfrak{R}(L)$, hence $\mathfrak{I}_i$ is finite. $M_1, M_2 \in \mathfrak{I}_i$ implies $M_1 \cap M_2 \in \mathfrak{I}_i$. Otherwise distributivity leads to a contradiction: $L_{i+1} = (M_1 \cap M_2) + L_{i+1} = (M_1 + L_{i+1}) \cap (M_2 + L_{i+1}) = L_i \cap L_i = L_i$, but L_{i+1} was $\mathfrak{L}$-maximal in L_i. Hence P_i is the minimal element in $\mathfrak{I}_i$ and lies therefore in $\mathfrak{R}(L)$. That P_i has $P_i \cap L_{i+1}$ as unique $\mathfrak{L}$-maximal sublattice is immediate from the definition of $\mathfrak{I}_i$. If conversely $P \in \mathfrak{R}_1(L)$, let $i \in \{1, \dots, t\}$ be the biggest index with

$P \subseteq L_i$. Then $P \in \mathfrak{I}_i$ and $P = P_i$. This proves the statement above. To obtain the desired basis choose a minimal subset $\mathfrak{L}_i$ of P_i such that $\overline{\mathfrak{L}}_i = \{x + (L_{i+1} \cap P_i) \mid x \in \mathfrak{L}_i\}$ is an $\Omega/\mathfrak{P}$-basis of $P_i/P_i \cap L_{i+1}$ for $i = 1,\ldots,t$. From Nakayama's Lemma it is clear that $\mathfrak{L} = \bigcup_{i=1}^{t} \mathfrak{L}_i$ is a basis of L (note, $\dim_D V =$ $= \dim_{\Omega/\mathfrak{P}} L/L\mathfrak{P} = \sum_{i=1}^{t} \dim_{\Omega/\mathfrak{P}} L_i/L_{i+1} = \sum_{i=1}^{t} \dim_{\Omega/\mathfrak{P}} P_i/P_i \cap L_{i+1}$.) If $L' \in \mathfrak{L}$, one obtains an Ω-basis of L' by multiplying all vectors in $\mathfrak{L}_i$ by an appropiate power of an ideal generator of $\mathfrak{P}$ for $i = 1,\ldots,t$, as can be seen from the above bijection between $\mathfrak{R}_1(L)$ and $\mathfrak{R}_1(L')$.

q.e.d.

Now the following characterization of graduated orders can be proved. Most of the equivalences have already been proved in [Ple 77] in the splitting field case (except (iii) and (iv)). In this generality the equivalences of (i) to (v) except (iii) are more or less explicite in [Ple 80a]. The equivalence of (i) and (ii), which is an immediate consequence of (II.7) has also been proved in [Rum 81b].

(II.8) Theorem. *Let* Λ *be an* R*-order in* $A = D^{n\times n}$ *and let* V *be an irreducible* A*-module. The following conditions are equivalent:*

(i) Λ *is a graduated order.*

(ii) Λ *is intersection of maximal orders in* A *and the* Λ*-lattices* $(\neq 0)$ *in* V *form a distributive lattice (with respect to* $+$ *and* $\cap$ *).*

(iii) Λ *is intersection of maximal orders in* A *and the twosided* Λ*-ideals form a distributive lattice.*

(iv) $\Lambda/Jac(\Lambda) \cong \bigoplus_{i=1}^{t} (\Omega/\mathfrak{P})^{n_i \times n_i}$ *with* $n_1+\ldots+n_t = n$ *and* Λ *is intersection of maximal orders in* A .

(v) For each irreducible Λ-lattice L the Λ-composition factors $S_1,\dots,S_t$ of the Λ-(torsion)module $L/L\mathfrak{P}$ all have multiplicity 1 in a Λ-composition series of $L/L\mathfrak{P}$, the Λ-endomorphism rings of $S_1,\dots,S_t$ are all isomorphic to $\Omega/\mathfrak{P}$ as $R/\mathfrak{p}$-algebras and Λ is intersection of maximal orders in A.

Proof: That (i) implies (ii), (iii), (iv), and (v) is a consequence of (II.3) and (II.4). (One can even obtain the graduated order Λ as the intersection of its maximal overorders corresponding to the projective indecomposable Λ-lattices.) That (ii) implies (i) follows from (II.7).

To prove the implication (iii) $\Rightarrow$ (ii), assume that the lattice $\mathfrak{Z}(V)$ of irreducible Λ-lattices in V is not distributive. Since $\mathfrak{Z}(V)$ is modular, there must be a sublattice of $\mathfrak{Z}(V)$ isomorphic to the subgroup lattice of the Klein four-group by a theorem by Birkhoff, cf. [Bir 67], i.e. there exist $L,L_1,L_2,L_3,L' \in \mathfrak{Z}(V)$ with $L_1 \neq L_2$, $L_1+L_2=L_1+L_3=L_2+L_3=L$, and $L_1 \cap L_3 = L_1 \cap L_3 = L_2 \cap L_3 = L'$. For $X,Y \in \mathfrak{Z}(V)$ let $I(X,Y)$ denote the twosided fractional ideal of of all $a \in A$ with $aX \subseteq Y$. The distributive law for sums and intersections of twosided ideals leads to the contradiction
$I(L,L_3)=(I(L,L_1) \cap I(L,L_2)) + I(L,L_3)=(I(L,L_1)+I(L,L_3)) \cap (I(L,L_2)+I(L,L_3))$
$= I(L,L)$. Note, for the last equality one uses $I(L,L_1)+I(L,L_3)=I(L,L)$ (and similarly for L_2 and L_3), which can be checked by a straightforward calculation with ideals of the order $I(L_1,L_1) \cap I(L_3,L_3)$, which is a graduated order by (II.7) (or the elementary divisor theorem).

To prove the implication (v) $\Rightarrow$ (iv) assume that there is a Λ-lattice L in the set $\mathfrak{Z}(V)$ of irreducible Λ-lattices in V such that $L\Omega \neq L$, i.e. the Λ-endomorphism ring of L is properly contained in Ω. Then

$L\Omega$ is a Λ-lattice, and there exists a $\varphi \in \Omega$ with $L\varphi \neq L$. But L and $L\varphi$ are Λ-sublattices of $L\Omega$, and φ induces a Λ-module isomorphism of $L\Omega/L$ onto $L\Omega/L\varphi$. One may assume that L is a maximal Λ-sublattice of $L\Omega$ (replace L by a maximal Λ-sublattice of $L\Omega$ containing L and choose a new $\varphi \in \Omega$ if necessary.) But then the simple Λ-module $L\Omega/L$ occurs at least twice as a composition factor in $L\Omega/L\Omega\mathfrak{P}$, contrary to the assumption in (v). Hence $L\Omega = L$ for all $L \in \mathfrak{J}(V)$. Therefore Ω maps into $\mathrm{End}_\Lambda(S_i)$ for $i = 1,\dots,t$, the images being isomorphic to $\Omega/\mathfrak{P}$. Because of the assumption in (v) the $\mathrm{End}_\Lambda(S_i)$ already coincide with these images. Let $n_i = \dim_{\Omega/\mathfrak{P}} S_i$. Then $\Lambda/\mathrm{Jac}(\Lambda) \cong \bigoplus_{i=1}^{t} (\Omega/\mathfrak{P})^{n_i \times n_i}$ follows. Since $n = \dim_{\Omega/\mathfrak{P}} L/L\mathfrak{P} = \sum_{i=1}^{k} n_i$ statement (iv) is proved.

Finally, to prove the implication (iv) $\Rightarrow$ (i), lift a complete set of orthogonal idempotents of $\Lambda/\mathrm{Jac}(\Lambda)$. Because of hypothesis (iv) this yields n orthogonal idempotents $\varepsilon_1,\dots,\varepsilon_n$ of Λ which sum up to 1. It has to be proved that $\varepsilon\Lambda\varepsilon$ is a maximal order in $\varepsilon A\varepsilon$ for each $\varepsilon \in \{\varepsilon_1,\dots,\varepsilon_n\}$. But since Λ is the intersection of maximal R-orders Γ in A, one obtains $\varepsilon\Lambda\varepsilon$ as intersection of the maximal R-orders $\varepsilon\Gamma\varepsilon$ in $\varepsilon A\varepsilon$ (note ε lies in Γ). But $\varepsilon A\varepsilon$ is isomorphic to the division algebra D, and hence $\varepsilon A\varepsilon$ contains exactly one maximal R-order. Therefore $\varepsilon\Lambda\varepsilon$ is a maximal order and Λ a graduated order.

q.e.d.

That Λ is the intersection of maximal orders is rather restrictive and usually not easy to check. To avoid this condition one is forced to introduce graduable orders as done in part b of this chapter. Note, however, if D is commutative, (II.8) (iv) and (v) can be modified by demanding that the center $Z(\Lambda)$ of Λ is a maximal order in $Z(A)$

instead of demanding that Λ is the intersection of its maximal overorders. At least under this additional hypothesis one then has a convenient criterion for an order to be a graduated order.

II.b Graduable orders

From now on it will always be assumed that the residue class field $R/\mathfrak{p}$ is <u>finite</u>. If R' is the maximal R-order in an extension field K' of K with $\dim_K K'$ <u>finite</u>, R' will be referred to as an extension of R. Graduable orders are R-orders which become graduated orders upon extending the ground ring R, cf. Definition (II.13) below. The proper tool for their investigation is the concept of a maximal separable suborder of an order, which will be treated first. Apart from their application to graduable orders, maximal separable suborders can be used to define the decomposition map of an order, as remarked by JACOBINSKI, cf. discussion following (III.10). They turn up already in [Ben 61] as maximal orders of the "inertia algebra".

<u>(II.9) Definition.</u> An R-order Σ is called separable (or an absolute maximal R-order), if the discriminant of Σ with respect to the reduced trace is equal to R.

The subsequent characterization of separable orders is well known, cf. e.g. [Rog 70].

<u>(II.10) Proposition.</u> For an R-order Σ the following properties are equivalent:

(i) Σ is separable;

(ii) $S \underset{R}{\otimes} \Sigma$ *is a maximal order for all extensions* S *of* R *;*

(iii) Σ *is the direct sum of full matrix rings over unramified extensions of* R *;*

(iv) $S \underset{R}{\otimes} \Sigma$ *is the direct sum of full matrix rings over* S *for a suitable unramified extension* S *of* R *.*

Proof: The implication (i) ⇒ (ii) is clear, since $\mathrm{discr}(S \underset{R}{\otimes} \Sigma) = S \cdot \mathrm{discr}(\Sigma)$.

To prove (ii) ⇒ (iii) note that Σ itself is a maximal order and hence the direct sum of full matrix rings over the maximal R-orders Ω_s of K-division algebras D_s. If Ω_s is not commutative, $S \underset{R}{\otimes} \Omega_s$ is only hereditary but not maximal in case S is chosen to be the maximal order of the inertia field of D_s. If Ω_s is commutative but ramified over R, choose $S = \Omega_s$ to find that $S \underset{R}{\otimes} \Omega_s$ is not maximal. Hence (ii) implies (iii).

The choice of S for a proof of the implication (iii) ⇒ (iv) is an unramified extension of R in which all the extensions mentioned in (iii) can be embedded.

Finally implication (iv) ⇒ (i) also follows from the formula $\mathrm{discr}(S \underset{R}{\otimes} \Sigma) = S \cdot \mathrm{discr}(\Sigma)$.

q.e.d.

<u>*(II.11) Corollary.*</u> *Let* Σ *be a separable R-order. Then* $\mathrm{Jac}(\Sigma) = \mathfrak{p}\Sigma$ *and the isomorphism type of* Σ *is determined by* $\Sigma/\mathrm{Jac}(\Sigma)$ *.*

Now let Λ be an R-order in the semisimple K-algebra A. A suborder Σ of Λ is understood to be an R-order contained in Λ with the same unit element as Λ (but not necessarily satisfying $K\Sigma = A$.) Clearly increasing chains of separable R-suborders of Λ terminate. Hence Λ contains maximal separable suborders. Their most important properties are given in the next proposition, cf. also [Ben 61]p. 130 for the conjugacy of inertia algebras.

(II.12) Proposition. *Let Λ be an arbitrary R-order.*

(i) If Σ is a maximal separable suborder of Λ , then $Jac(\Sigma) = \Sigma \cap Jac(\Lambda)$ and $\Lambda/Jac(\Lambda) \cong \Sigma/Jac(\Sigma)$.

(ii) Any two maximal separable suborders of Λ can be transformed into each other by units of Λ .

Proof: (i) For any separable suborder Σ of Λ one certainly has $Jac(\Lambda) \cap \Sigma \subseteq Jac(\Sigma)$. Since $Jac(\Sigma) = \mathfrak{p}\Sigma \subseteq \mathfrak{p}\Lambda \subseteq Jac(\Lambda)$ equality holds. It is convenient to prove the second statement of (i) in several steps. Note, by (II.11) a separable suborder Σ of Λ is a maximal separable suborder, if $\Sigma/Jac(\Sigma) \cong \Lambda/Jac(\Lambda)$.

Step 1. For any order Λ there exists a separable suborder Σ with $\Sigma/Jac(\Sigma) \cong \Lambda/Jac(\Lambda)$.

For the proof of this statement it is without loss of generality to assume Λ to be local because of the possibility of lifting matrix units. The unit group U_i of $\Lambda/\mathfrak{p}^i\Lambda$ contains a normal p-subgroup N_i , $p = char(R/\mathfrak{p})$, namely $N_i = \{x + \mathfrak{p}^i\Lambda \mid x \in 1 + Jac\Lambda\}$ such that U_i/N_i is a cyclic group of order p^k-1 with $k = \dim_{\mathbb{Z}/p\mathbb{Z}}\Lambda/Jac(\Lambda)$. By the Schur-Zassenhaus-Theorem U_i splits over N_i , i.e. there exists an $x_i \in \Lambda$ such that x_i is a primitive (p^k-1)-th root of unity in $\Lambda/\mathfrak{p}^i\Lambda$. Since Λ is compact in the $\mathfrak{p}$-adic topology, the sequence $(x_i)_{i\in\mathbb{N}}$ has a convergent subsequence converging to some $x \in \Lambda$. (Actually, a slight refinement of the above argument shows that

$(x_i)_{i\in\mathbb{N}}$ can be chosen as converging sequence.) $\Sigma = R[x]$ has the desired property.

<u>Step 2.</u> For any two separable suborders Σ_1 and Σ_2 of an arbitrary R-order Λ with $\Sigma_1/\mathrm{Jac}(\Sigma_1) \cong \Sigma_2/\mathrm{Jac}(\Sigma_2) \cong \Lambda/\mathrm{Jac}(\Lambda)$ there exists a unit x of Λ such that $x^{-1}\Sigma_1 x = \Sigma_2$.

Since matrix units which are conjugate mod $\mathrm{Jac}(\Lambda)$ are already conjugate in Λ , cf. [Zas 54], it suffices to assume that Λ is local. In the notation of the proof of Step 1 above, N_i has two complements $C_i^{(\alpha)} = \{x + \mathfrak{p}^i\Lambda \mid x \text{ unit in } \Sigma_\alpha\}, \alpha = 1,2,$ in the unit group U_i of $\Lambda/\mathfrak{p}^i\Lambda$ According to the Schur-Zassenhaus- Theorem there is a unit $x_i + \mathfrak{p}^i\Lambda$ in U_i transforming $C_i^{(1)}$ onto $C_i^{(2)}$. The desired unit x can be chosen such that a suitable subsequence of $(x_i)_{i\in\mathbb{N}}$ converges to x as in Step 1.

<u>Step 3.</u> For a commutative separable suborder Σ' of Λ there exists a separable suborder Σ of Λ containing Σ' and satisfying $\Sigma/\mathrm{Jac}(\Sigma) \cong \Lambda/\mathrm{Jac}(\Lambda)$.

By Step 1 there exists a separable suborder $\tilde{\Sigma}$ of Λ with $\tilde{\Sigma}/\mathfrak{p}\tilde{\Sigma} = \Lambda/\mathrm{Jac}(\Lambda)$. Let Σ'' be a separable suborder of $\tilde{\Sigma}$ with $\Sigma'' + \mathrm{Jac}(\Lambda)/\mathrm{Jac}(\Lambda) = \Sigma' + \mathrm{Jac}(\Lambda)/\mathrm{Jac}(\Lambda)$. It suffices to prove that $x^{-1}\Sigma'' x = \Sigma'$ for a suitable unit x of Λ . This follows by the same argument as in Step 2 (with Λ replaced by $\Sigma' + \mathrm{Jac}(\Lambda)$).

<u>Step 4.</u> If Σ' is a separable suborder of Λ such that $\Sigma'/\mathfrak{p}\Sigma'$ and $\Lambda/\mathrm{Jac}(\Lambda)$ are simple $R/\mathfrak{p}$-algebras, then there exists a separable suborder Σ of Λ containing Σ' with $\Sigma/\mathfrak{p}\Sigma \cong \Lambda/\mathrm{Jac}(\Lambda)$.

Let ε be a primitive idempotent of Σ' . Then Σ' and Λ can be identified with $(\varepsilon\Sigma'\varepsilon)^{n\times n}$ and $(\varepsilon\Lambda\varepsilon)^{n\times n}$ for a suitable natural number n . Since $\varepsilon\Sigma'\varepsilon$ is a commutative separable suborder of $\varepsilon\Lambda\varepsilon$, Step 3 can be applied.

Step 5. If Σ' is any separable suborder of Λ , then Λ contains a separable suborder $\Sigma \supseteq \Sigma'$ with $\Sigma/\mathfrak{p}\Sigma \cong \Lambda/\mathrm{Jac}(\Lambda)$.

For the proof it suffices to assume $\varepsilon\Sigma'\varepsilon/\mathrm{Jac}(\varepsilon\Sigma'\varepsilon) \cong \varepsilon\Lambda\varepsilon/\mathrm{Jac}(\varepsilon\Lambda\varepsilon)$ for all primitive central idempotents ε of Σ' because of Step 3 and Step 4. Moreover, by passing from Λ to $\varepsilon\Lambda\varepsilon$, where ε is the sum of all primitive central idempotents of Σ' whose residue classes modulo $\mathrm{Jac}(\Lambda)$ lie in the same component of $\Lambda/\mathrm{Jac}(\Lambda)$, one can assume that $\Lambda/\mathrm{Jac}(\Lambda)$ is simple. By standard arguments Λ has a complete set of matrix units such that a unique subset of this set generates Σ' over the center of Σ' . Let $\tilde{\varepsilon}$ be a primitive idempotent of Σ' , $\tilde{\Sigma}' = \tilde{\varepsilon}\Sigma'\tilde{\varepsilon}$, and $\tilde{\Lambda} = \tilde{\varepsilon}\Lambda\tilde{\varepsilon}$. Applying Step 2 to $\varepsilon\Lambda\varepsilon$ for each primitive central idempotent ε of Σ' , yields an identification of $\varepsilon\Sigma'\varepsilon$ with $(\tilde{\Sigma}')^{n(\varepsilon)\times n(\varepsilon)} \subseteq \tilde{\Lambda}^{n(\varepsilon)\times n(\varepsilon)}$ the latter matrix ring being identified with $\varepsilon\Lambda\varepsilon$, where $n(\varepsilon)$ is a natural number depending on ε . Moreover, in the same way Λ is identified with $\tilde{\Lambda}^{n\times n}$ where n is the sum of the $n(\varepsilon)$ just defined and the direct sum of the $\varepsilon\Lambda\varepsilon$ corresponds to block diagonal matrices over $\tilde{\Lambda}$ with block sizes $n(\varepsilon)$. The desired separable suborder Σ of Λ then corresponds to $(\tilde{\Sigma}')^{n\times n}$. This finishes the proof of (II.12)(i).

(ii) This statement is now an immediate consequence of statement (i) and Step 2 of the proof of (i).

q.e.d.

The technical apparatus for the investigation of graduable orders is now ready.

(II.13) Definition. *An R-order Λ in the semisimple K-algebra A is called graduable, if there exists an unramified extension S of R such that $S \otimes_R \Lambda$ is a graduated S-order.*

As an immediate consequence of the equality of the center $Z(S \otimes_R \Lambda)$ of $S \otimes_R \Lambda$ with $S \otimes_R Z(\Lambda)$ one gets the next remark.

(II.14) Remark. *The center $Z(\Lambda)$ of a graduable order Λ is the maximal order of the center $Z(A)$ of the algebra A. In particular, if A_s, $s = 1,\dots,h$, are the simple components of A, then $\Lambda = \bigoplus_{s=1}^{h} \Lambda_s$ with $\Lambda_s = A_s \cap \Lambda$.*

Because of this the investigation of graduable orders can be restricted to orders in central simple algebras $A = D^{n\times n}$. The next result characterizes graduable orders by means of their maximal separable suborders.

(II.15) Proposition. *Let Λ be an R-order in the central simple K-algebra $A = D^{n\times n}$ and let Σ be a maximal separable suborder of Λ. The following statements are equivalent.*

(i) Λ is a graduable R-order.

(ii) The Σ-Σ-lattices I in A with $KI = A$ form a distributive lattice (with respect to $+$ and $\cap$).

(iii) Λ decomposes as Σ-Σ-lattice into the direct sum of pairwise non-isomorphic irreducible Σ-Σ-sublattices.

(iv) For any maximal R-order Γ containing Λ the simple Γ-module X decomposes into the sum of simple non-isomorphic Σ-modules, if X is viewed as Σ-module.

Proof. For the proof of implication (i) $\Rightarrow$ (ii) note that the mapping

$I \to S \otimes_R I$ of set $\mathfrak{L}$ of Σ-Σ-lattices I in A with $KI = A$ into the corresponding set $\tilde{\mathfrak{L}}$ of $(S \otimes \Sigma) - (S \otimes \Sigma)$-sublattices of $\tilde{K} \otimes_K A$, $\tilde{K} = S \otimes_R K$, is injective for each (unramified) extension S of R. The possibility of choosing compatible R-bases for any two Σ-Σ-lattices in $\mathfrak{L}$, cf. (II.7), shows that this map is a monomorphism of $(+,\cap)$-lattices. If S is chosen big enough such that $S \otimes_R \Lambda$ is a graduated S-order, then it is clear that the lattice $\tilde{\mathfrak{L}}$ is distributive, hence also the lattice $\mathfrak{L}$ given in (ii) since it can be viewed as sublattice of $\tilde{\mathfrak{L}}$.

To prove (iii) under the hypothesis of (ii) note that $\Sigma \otimes_R \Sigma^{op}$ is again a separable order. Hence Λ decomposes as Σ-Σ-lattice into the direct sum of irreducible Σ-Σ-sublattices. If two of them were isomorphic one immediately gets a contradiction to the distributivity in (ii) (see proof of implication (iii)$\Rightarrow$(ii) of (II.8) for the configuration leading to a contradiction).

The proof of implication (iii) $\Rightarrow$ (iv) follows from the observation that $\Gamma/\mathrm{Jac}(\Gamma)$ decomposes under the hypothesis of (iii) into the direct sum of pairwise non-isomorphic simple Σ-Σ-modules. This characterizes the centralizer of $\Sigma + \mathrm{Jac}(\Gamma)/\mathrm{Jac}(\Gamma)$ in $\Gamma/\mathrm{Jac}(\Gamma)$ to the extent that (iv) follows.

It remains to prove implication (iv) $\Rightarrow$ (i). Because of Definition (II.1) of a graduated order one may assume Σ to be local. Under this additional hypothesis $\Sigma + \mathrm{Jac}(\Gamma)/\mathrm{Jac}(\Gamma)$ is a selfcentralizing subfield of $\Gamma/\mathrm{Jac}(\Gamma)$. Hence $\dim_{R/\mathfrak{p}} \Sigma/\mathfrak{p}\Sigma = n \cdot \dim_{R/\mathfrak{p}} \Omega/\mathfrak{P}$. Call this number m; then $\Sigma \otimes_R \Sigma \cong \bigoplus_{i=1}^{m} \Sigma$ and $\Sigma \otimes_R \Lambda$ is a graduated Σ-order.

q.e.d.

Before a rather explicite description of graduable orders (based on (II.15)(iii)) will be derived it might be of general interest to have a characterization of graduable orders analogous to that of graduated orders in (II.8) which - unlike to (II.15) - is not based on the interplay with a maximal separable suborder. The numbering of the statement in Theorem (II.16) is supposed to simplify the comparison with (II.8). There is no counterpart to (II.8)(ii).

<u>(II.16) Theorem.</u> *Let* Λ *be an* R*-order in the central simple* K*-algebra* $A = D^{n\times n}$ *. The following conditions are equivalent.*

(i') Λ is a graduable R-order.

(iii') The twosided ideals of Λ form a distributive lattice (with respect to $+$ and $\cap$).

(iv') $\Lambda/Jac(\Lambda) \cong \bigoplus_{i=1}^{t} F_i^{n_i\times n_i}$, where F_i are finite extension fields of $F = \Omega/\mathfrak{P}$, and $n_1 dim_F F_1 + \ldots + n_t dim_F F_t = n$.

(v') For any irreducible Λ-lattice L with $L\Omega = L$ each Λ-composition factor of $L/L\mathfrak{P}$ has multiplicity 1 in a composition series of $L/L\mathfrak{P}$.

Proof. Implications (i') $\Rightarrow$ (iii'), (i') $\Leftrightarrow$ (v'), and (iv') $\Rightarrow$ (i') follow from implications (i) $\Rightarrow$ (ii), (i) $\Leftrightarrow$ (iv) of (II.15), and Definition (II.13) respectively. It remains to prove that (iii') implies (iv'). Under hypothesis (iii') the intersection $\overline{\Lambda}$ of all maximal R-orders in A containing Λ is a graduated R-order by (II.8). If V is an irreducible A-module, the $\overline{\Lambda}$-lattices in V coincide with the Λ-Ω-lattices in V . Let $L_1 > L_2 > \ldots > L_m = L_1\mathfrak{P}$ be a chain of $\overline{\Lambda}$-lattices in V with L_i maximal in L_{i+1} . Then $Jac(\overline{\Lambda}) = \{x \in \overline{\Lambda} \mid x L_i \subseteq L_{i+1} , i = 1,\ldots,m-1\}$ and, since L_i/L_{i+1} splits into a direct sum of simple isomorphic Λ-modules, $Jac(\Lambda) = \{x \in \Lambda \mid x L_i \subseteq L_{i+1} , i = 1,\ldots,m-1\}$. Hence $Jac(\Lambda) = \Lambda \cap Jac(\overline{\Lambda})$.

If $\overline{\Lambda}=\Lambda$, (iv') is correct by (II.8). If $\overline{\Lambda}\neq\Lambda$, the distributivity of the intervall of twosided Λ-ideals between $\overline{\Lambda}$ and $\mathrm{Jac}(\overline{\Lambda})$, forces $\Lambda/\mathrm{Jac}(\Lambda)$ to have as many components as $\overline{\Lambda}/\mathrm{Jac}(\overline{\Lambda})$ (as $(R/\mathfrak{p})$-algebras) and to compensate the (smaller) matrix degrees of the components of $\Lambda/\mathrm{Jac}(\Lambda)$ by the degrees of the F_i over $R/\mathfrak{p}$, cf. also proof of (iii) $\Rightarrow$ (iv) of (II.15).

q.e.d.

Local graduable orders are special cases of the crossed product orders, investigated by BENZ and ZASSENHAUS in [BeZ 83], they are also graded rings in the sense of [NaO 82].

(II.17) Proposition. Let Λ *be a local R-order in the central simple K-algebra* A, Σ *a maximal separable suborder of* Λ, *and* $n=\dim_{R/\mathfrak{p}}\Lambda/\mathrm{Jac}(\Lambda)$. *Then*

(i) Λ *splits as* Σ-Σ-*lattice into the direct sum of the* n *maximal irreducible* Σ-Σ-*sublattices* $\Lambda(0),\ldots,\Lambda(n-1)$ *defined by*

$$\Lambda(i)=\{x\in\Lambda \mid yx=x\varphi^i(y) \quad \text{for all} \quad y\in\Sigma\} \qquad (0\leq i<n)$$

where φ *denotes the Frobenius automorphism of the extension* Σ/R.

(ii) There are numbers $\alpha(i,j)\in\mathbb{Z}_{\geq 0}$, $0\leq i,j<n$ *such that*

$$\Lambda(i)\Lambda(j)=\mathfrak{p}^{\alpha(i,j)}\Lambda(i+j)$$

(with $i+j$ *interpreted modulo* n *).* α *can be viewed as 2-cocycle of the cyclic group* $\mathbb{Z}/n\mathbb{Z}$ *taking values in the trivial module* $\mathbb{Z}$.

(iii) The Σ-Σ-*lattices* L *in* A *with* $KL=A$ *are in 1-1-correspondence with the* $\mathbb{Z}$-*valued* $\mathbb{Z}/n\mathbb{Z}$-*cocycles which differ from* α *(defined in (ii)) by coboundaries. The resulting cohomology class in* $H^2(\mathbb{Z}/n\mathbb{Z},\mathbb{Z})$ *determines the isomorphism type of the K-algebra* A *: The Hasse invariant of* A *is congruent to* $\frac{s}{n}$

mod $\mathbf{Z}$ *with* $s=\tilde{\alpha}(1,1)+\tilde{\alpha}(1,2)+\ldots+\tilde{\alpha}(1,n)$ *for any cocycle* $\tilde{\alpha}$ *representing the class. The* Σ-Σ-*lattice* L *is an R-order in* A *if and only if the associated cocycle does not take negative values.*

(iv) *The local R-order* Λ *(as well as* $A = K\Lambda$ *) is uniquely determined by* n *and the cocycle* α *defined in (ii).*

(v) *Each 2-cocycle* $\alpha : \mathbf{Z}/n\mathbf{Z}\times\mathbf{Z}/n\mathbf{Z} \to \mathbf{Z}$ *, which takes nonnegative values and satisfies* $\alpha(0,0)=0$ *, defines a local R-order* Λ_α *in a central simple K-algebra* A_α *(* A_α *depends only on the cohomology class of* α *) with* Σ *as a separable suborder and* $\dim_{R/\mathfrak{p}}\Lambda_\alpha/Jac(\Lambda_\alpha) = n^2/\min\{k\in\mathbf{N} \mid k|n \text{ , } \alpha(k,k)+\alpha(k,2k)+\ldots+\alpha(k,n)=0\}$.

Proof: (i) The first statement is a consequence of (II.15)(iii). That none of the $\Lambda(i)$ described in (i) is equal to 0 follows from the Skolem-Noether-Theorem.

(ii) The description of the $\Lambda(i)$ in (i) implies $\Lambda(i)\Lambda(j)\subseteq\Lambda(i+j)$. Since $\Lambda(i+j)$ is an irreducible Σ-Σ-lattice and Σ a separable R-order, one obtaines $\Lambda(i)\Lambda(j) = \mathfrak{p}^{\alpha(i,j)}\Lambda(i+j)$ for a suitable $\alpha(i,j)\in\mathbf{Z}_{\geq 0}$. The cocycle condition for α follows from associativity: $(\Lambda(i)\Lambda(j))\Lambda(k) = \Lambda(i)(\Lambda(j)\Lambda(k))$ for $0\leq i,j,k<n$.

(iii) Again by (II.15)(iii) there exist $\beta(i)\in\mathbf{Z}$, $0\leq i<n$, depending on the Σ-Σ-lattice L such that $L = \bigoplus_{i=0}^{n-1}\mathfrak{p}^{\beta(i)}\Lambda(i)$. Analogous to the definition of α in (ii) one gets the associated cocyle as $\alpha_L(i,j) = \alpha(i,j)+\beta(i)+\beta(j)-\beta(i+j)$. Clearly L is multiplicatively closed iff $\alpha(i,j)\geq 0$ for all $i,j\in\mathbf{Z}/n\mathbf{Z}$. In this situation L contains 1 , iff $\alpha(0,0)=0$. To compute the Hasse invariant of A , note that A is a crossed product algebra of $K\Sigma$ with $Gal(K\Sigma/K)\cong\mathbf{Z}/n\mathbf{Z}$. Let $a\in\mathfrak{p}^{\beta(1)}\Lambda(1)$ contained in some Σ-Σ-lattice L , then $a^n\in\mathfrak{p}^s\Sigma\setminus\mathfrak{p}^{s+1}\Sigma$ with $s=\alpha_L(1,1)+\alpha_L(1,2)+\ldots+\alpha_L(1,n)$. This

implies the statement about the Hasse invariant, cf. e.g. [Rei 75] p. 265.

(iv) This follows from (iii), the Skolem-Noether-Theorem, and (i) and (ii).

(v) Let $\tilde{K}$ be an unramified extension of K of degree n , Σ the maximal R-order of $\tilde{K}$, define s by $s = \alpha(1,1)+\alpha(1,2)+\ldots+\alpha(1,n)$, and let $\varphi \in Gal(\tilde{K}/K)$ induce the Frobenius automorphism of Σ . The Σ-Σ-lattices and the R-orders containing Σ in the crossed product algebra $(\tilde{K}/K,\varphi,a)$ with $a \in \mathfrak{p}^s\Sigma \setminus \mathfrak{p}^{s+1}\Sigma$ can be described by cocycles cohomologous to α . (Note, $H^2(\mathbb{Z}/n\mathbb{Z},\mathbb{Z})$ is cyclic of order n .) This associates a unique R-order Λ_α with α . Checking for which i , $1 \leq i \leq n-1$, $\Lambda(i)^n$ is contained in $\mathfrak{p}\Sigma$ yields the dimension of $\Lambda_\alpha/Jac(\Lambda_\alpha)$.

q.e.d.

For a non-local graduable order Λ one can also give a detailed description generalizing both, the description of graduated orders by their structural invariants discussed in Part a of this chapter and the cocycle description of local graduable orders in (II.17). This is only pursued to the extent needed for (II.20). For the investigation one may assume without loss of generality that $\Lambda/Jac(\Lambda)$ is abelian, since each graduable order is Morita equivalent to a graduable order with this property. The next definition simplifies the notation.

<u>(II.18) Definition.</u> *Let* $\Sigma = \bigoplus_{i=1}^{t} \Sigma_i$ *be the decomposition of a separable commutative R-order* Σ *into components* Σ_i *with* $d_i = dim_R \Sigma_i$. *Let* $\tilde{\Sigma}$ *be the unramified extension of degree* $d = gcm(d_1,\ldots,d_t)$ *of* R , *choose embeddings* $\iota_i : \Sigma_i \to \tilde{\Sigma}$, *and set* $\tilde{\Sigma}_i = \iota_i(\Sigma_i)$. *The <u>frame</u> of* Σ

with respect to the ι_i *is the system of R-suborders* $\Sigma_{ij} = \iota_i^{-1}(\tilde{\Sigma}_i \cap \tilde{\Sigma}_j)$ *together with the R-order isomorphisms* $\varphi_{ij}^{(\nu)} : \Sigma_{ij} \to \Sigma_{ji}$ *corresponding to the* ν*-th power of the Frobenius on* $\tilde{\Sigma}_i \cap \tilde{\Sigma}_j$, $\nu \in \mathbb{Z}$, $1 \le i,j \le t$.

(II.19) Proposition. *Let* Λ *be a graduable R-order in the central simple K-algebra* A *with* $\Lambda/Jac(\Lambda)$ *commutative. Fix a maximal separable suborder* $\Sigma = \bigoplus_{i=1}^{t} \Sigma_i$ *of* Λ *with components* Σ_i , $1 \le i \le t$ *and fix a frame* $(\Sigma_{ij}, \varphi_{ij}^{(\nu)})$ *of* Σ . *Let* $d_i = dim_R \Sigma_i$, $d_{ij} = ggT(d_i, d_j)$, *and* $d_{ijk} = ggT(d_i, d_j, d_k)$ *for* $1 \le i,j,k \le t$.
Then

(i) Λ *splits as* Σ-Σ*-lattice into the direct sum of the maximal irreducible* Σ-Σ*-sublattices* $\Lambda_{ij}(\nu)$ *defined by*
$\Lambda_{ij}(\nu) = \{x \in \Sigma_i \Lambda \Sigma_j \mid y x = x \varphi_{ij}^{(\nu)}(y)$ *for all* $y \in \Sigma_{ij}\}$ *for* $1 \le i,j \le t$, $0 \le \nu < d_{ij}$

(ii) *There exist numbers* $m_{ijk}(\lambda,\mu,\nu) \in \mathbb{Z}_{\ge 0}$ *for* $1 \le i,j,k \le t$, $0 \le \lambda < d_{ij}$, $0 \le \mu < d_{ik}$, $0 \le \nu < d_{ik}$ *and* $\nu \equiv \mu + \nu \pmod{d_{ijk}}$ *such that*
$$\Lambda_{ij}(\lambda)\Lambda_{jk}(\mu) = \bigoplus_{\nu} p^{m_{ijk}(\lambda,\mu,\nu)} \Lambda_{ik}(\nu)$$
where the ν*'s are restricted as just defined.*

(iii) *Let* Λ' *be a second graduable R-order in* A *containing* Σ *as a maximal seperable suborder. If* Λ' *leads to the same constants* $m_{ijk}(\lambda,\mu,\nu)$, *then* Λ' *is isomorphic to* Λ *under an isomorphism induced by conjugation by a unit of* $K\Sigma$.

Proof: (i) By (II.15) (iii) Λ splits into the direct sum of irreducible Σ-Σ-lattices. That the irreducible Σ-Σ-lattices in $\Sigma_i \Lambda \Sigma_i$ are the $\Lambda_{ii}(\nu)$ described above follows from (II.17)(i). It remains to prove that $\Lambda_{ij}(\nu)$ is an irreducible Σ-Σ-lattice $(\neq 0)$ and $\Sigma_i \Lambda \Sigma_j = \bigoplus_{\nu=0}^{d_{ij}-1} \Lambda_{ij}(\nu)$ for fixed i,j with $i \neq j$. Since A is simple,

$\Sigma_i \Lambda \Sigma_j \neq 0$. Clearly $\Lambda_{ij}(\nu)$ is a Σ-Σ-sublattice of $\Sigma_i \Lambda \Sigma_j$. Hence at least one $\Lambda_{ij}(\nu_o) \neq 0$. But then all $\Lambda_{ij}(\nu) \neq 0$ since $0 \neq \Lambda_{ii}(\nu-\nu_o)\Lambda_{ij}(\nu_o) \subseteq \Lambda_{ij}(\nu)$ (read $\nu-\nu_o$ modulo d_i). On the other hand $\Sigma_i \circledast_R \Sigma_j^{op}$ splits into exactly d_{ij} components. Hence each $\Lambda_{ij}(\nu)$ is irreducible for $\nu = 0,\ldots,d_{ij}-1$ and the claim follows.

(ii) $\Lambda_{ij}(\lambda)\ \Lambda_{jk}(\mu)$ is a non-zero Σ-Σ-sublattice of the Σ-Σ-lattice

$$\{x \in \Sigma_i \Lambda \Sigma_k \mid y\, x = x\ \varphi_{ik}^{(\lambda+\mu)}(y) \quad \text{for all} \quad y \in \Sigma_{ij} \cap \Sigma_{ik}\} =$$
$$= \bigoplus_{\nu \in I} \Lambda_{ik}(\nu) \quad \text{with} \quad I = \{\nu \mid 0 \leq \nu < d_{ij}\ ,\ \nu \equiv \lambda+\mu \mod d_{ijk}\}\ .$$

Hence $\Lambda_{ij}(\lambda)\Lambda_{jk}(\mu) = \bigoplus_{\nu \in I} X_\nu$ where X_ν is a Σ-Σ-sublattice of $\Lambda_{ik}(\nu)$. It remains to prove $X_\nu \neq 0$ for all $\nu \in I$. To this end recall the notation of (II.18). Since the field of quotients of $\tilde{\Sigma}$ splits A , the R-order Λ can be embedded into $\tilde{\Sigma}^{d \times d}$ with $d = d_1+\ldots+d_t$. This can be done in such a way that the elements of Σ_i are mapped onto the diagonal matrices diag $(0,\ldots,0,x,\varphi(x),\ldots \varphi^{d_i-1}(x),0,\ldots,0)$ with $x \in \tilde{\Sigma}_i$ in the $(d_1+\ldots+d_{i-1}+1)$st position, where φ denotes the Frobenius of $\tilde{\Sigma}$ over R . Clearly, $\Sigma_i \Lambda \Sigma_j$ is mapped into the set of block matrices $(x_{kl}) \in \tilde{\Sigma}^{d \times d}$ with $x_{kl} \in \tilde{\Sigma}^{d_k \times d_l}$ equal to zero unless $(k,l) = (i,j)$, $(1 \leq k,l \leq t)$. More precisely, the non-zero elements of $\Lambda_{ij}(\nu)$ are mapped onto certain of these matrices (x_{kl}) where additionally the (i,j)-block $x_{ij} = (y_{rs}) \in \tilde{\Sigma}^{d_i \times d_j}$ satisfies $y_{rs} = 0$ iff $s-r \not\equiv \nu (\mathrm{mod}\ d_{ij})$. With this description the above claim follows easily.

(iii) Define $\Lambda'_{ij}(\nu)$ and $m'_{ijk}(\lambda,\mu,\nu)$ analogous to $\Lambda_{ij}(\nu)$ and $m_{ijk}(\lambda,\mu,\nu)$. Then there exist $m_{ij}(\nu) \in \mathbf{Z}$ with $\Lambda'_{ij}(\nu) = \mathfrak{p}^{m_{ij}(\nu)}\ \Lambda_{ij}(\nu)$ for $1 \leq i,j \leq t$ and $0 \leq \nu < d_{ij}$. Proposition (II.17) implies $\Lambda'_{ii}(\nu) = \Lambda_{ii}(\nu)$ for $1 \leq i \leq t$ and $0 \leq \nu < d_i$. Hence $\Lambda'_{ii}(\lambda)\Lambda'_{ij}(\mu) = \mathfrak{p}^{m_{ij}(\nu)}\ \Lambda_{ii}(\lambda)\ \Lambda_{ij}(\mu)$. Part (ii) together with $m'_{ijk}(\lambda,\mu,\nu) = m_{ijk}(\lambda,\mu,\nu)$ implies now that $m_{ij}(\nu)$ does not depend on ν . Set $m_{ij} = m_{ij}(\nu)$. A similar argument shows $m_{ij} + m_{jk} - m_{ik} = 0$ for

$1 \leq i,j,k \leq t$. Hence Lemma (II.5) implies the existence of $m_1,\dots,m_t \in \mathbb{Z}$ with $m_{ij} = m_i - m_j$. The claim follows immediately.

q.e.d.

The following theorem reduces many problems about graduable orders to questions about graduated orders. It is because of this that mainly graduated instead of graduable orders can and will be used in Chapters IIc, III and IV.

(II.20) Theorem. Let Λ *be a graduable R-order in the central simple K-Algebra* A *and let* S *be an unramified extension of* R *. Then the isomorphism type of* Λ *is uniquely determined by* $\bar{\Lambda} = S \otimes_R \Lambda$ *and the embedding* $\tau: \Lambda/Jac(\Lambda) \hookrightarrow \bar{\Lambda}/Jac(\bar{\Lambda})$ *induced by the natural embedding* $\Lambda \hookrightarrow \bar{\Lambda}$ *.*

Proof. Since Λ may be replaced by a suitable Morita-equivalent R-order it is without loss of generality to assume that $\Lambda/\mathrm{Jac}(\Lambda)$ is commutative. The given information about Λ is encoded in the diagram $\bar{\Lambda} \xrightarrow{\sigma} \bar{\Lambda}/\mathrm{Jac}(\bar{\Lambda}) \xleftarrow{\tau} \Lambda/\mathrm{Jac}(\Lambda)$ where σ is the natural epimorphism. Let Λ' be a second graduable R-order, $\bar{\Lambda}' \xrightarrow{\sigma'} \bar{\Lambda}'/\mathrm{Jac}(\bar{\Lambda}') \xleftarrow{\tau'} \Lambda'/\mathrm{Jac}(\Lambda')$ the corresponding diagram for Λ' such that the following diagram is commutative,

$$\begin{array}{ccccc} \bar{\Lambda} & \xrightarrow{\sigma} & \bar{\Lambda}/\mathrm{Jac}(\bar{\Lambda}) & \xleftarrow{\tau} & \Lambda/\mathrm{Jac}(\Lambda) \\ \chi \uparrow & & \uparrow \psi & & \omega \uparrow \\ \bar{\Lambda}' & \xrightarrow{\sigma'} & \bar{\Lambda}'/\mathrm{Jac}(\bar{\Lambda}') & \xleftarrow{\tau'} & \Lambda'/\mathrm{Jac}(\Lambda') \end{array}$$

where χ , ψ , ω resp. are S-order, $S/\mathrm{Jac}(S)$-, resp. $R/\mathfrak{p}$-algebra isomorphisms. The claim is that Λ and Λ' are isomorphic R-orders.

Let $\Sigma = \bigoplus_{i=1}^{t} \Sigma_i$ be a maximal separable suborder of Λ and $(\Sigma_{ij}, \varphi_{ij}^{(\nu)})$

be the frame of Σ with respect to fixed embeddings $\iota_i : \Sigma_i \to \widetilde{\Sigma}$ in the terminology of (II.18). Since ω is an isomorphism, a maximal separable suborder Σ' of Λ' is isomorphic to Σ. More precisely, $\Sigma' = \bigoplus_{i=1}^{t} \Sigma_i'$ and for each i, $1 \leq i \leq t$, there is an R-order isomorphism $\iota_i' : \Sigma_i' \to \Sigma_i$ compatible with ω. Let $(\Sigma_{ij}', \varphi_{ij}'^{(\nu)})$ be the frame of Σ' with respect to the embeddings $\iota_i \iota_i' : \Sigma_i' \to \widetilde{\Sigma}$. Define $\Lambda_{ij}(\nu)$, $m_{ijk}(\lambda,\mu,\nu)$ and $\Lambda_{ij}'(\nu)$, $m_{ijk}'(\lambda,\mu,\nu)$ as in (II.19). Note, $\overline{\Sigma} = S \otimes_R \Sigma$ is a maximal separable S-suborder of $\overline{\Lambda}$. Since τ and σ are known, $\overline{\Lambda}_{ij}(\nu) = S \otimes_R \Lambda_{ij}(\nu)$ can be identified as $\overline{\Sigma}$-$\overline{\Sigma}$-sublattice of $\overline{\Lambda}$. Hence the $m_{ijk}(\lambda,\mu,\nu)$ can be computed from the structural invariants of $\overline{\Lambda}$. The $m_{ijk}'(\lambda,\mu,\nu)$ are obtained from $\overline{\Lambda}'$ via σ' and τ' in the same way. Because of the commutativity of the above diagram and the compatible choices of frames for Σ and Σ', one obtains $m_{ijk}'(\lambda,\mu,\nu) = m_{ijk}(\lambda,\mu,\nu)$ for all parameters $i, j, k, \lambda, \mu, \nu$.

Since the $\alpha_i(\lambda,\mu) := m_{iii}(\lambda,\mu,\lambda+\mu)$ can be interpreted as cocycle describing both local R-orders, $\Sigma_i \Lambda \Sigma_i$ and $\Sigma_i' \Lambda' \Sigma_i'$, in the sense of (II.17) one may assume $A' = A$ and $\Lambda_{ii}'(\nu) = \Lambda_{ii}(\nu)$. In particular $\Sigma = \Sigma'$. Now the claim follows from (II.19).

q.e.d.

II.c. Some properties of graduated orders, graduated hulls

In view of the applications in the subsequent chapters some special facts about graduated orders are needed. The notation of Chapter II.a is kept. The first concernes the codifferent or dual (resp. "inverse") different $\mathfrak{D}(\Lambda)$ of the graduated order Λ in the separable K-algebra $A = D^{n \times n}$. Let $tr_{A/K}$ denote the relative reduced trace of A, i.e. the composition of the reduced trace of A to the center $Z(A)$

of A and the trace map of $Z(A)$ to K, cf. [Rei 75]. For any R-order Λ the codifferent $\mathfrak{D}(\Lambda) = \mathfrak{D}_{A/K}(\Lambda)$ is the twosided fractional Λ-ideal of all $x \in A$ with $tr_{A/K}(x\Lambda) \subseteq R$, cf. [Ser 79], [Jac 81].

(II.21) Proposition. *Let the residue class field $R/\mathfrak{p}$ of R be finite, denote by m the index of D over its center Z (i.e. $\dim_Z D = m^2$), let $\bar{\mathfrak{p}}$ be the maximal ideal of the integral closure $\bar{R}$ of R in Z, and let $\bar{\mathfrak{p}}^{d}$ be the different of $\bar{R}$ over R. The codifferent $\mathfrak{D}_{A/K}(\Lambda)$ of the graduated order $\Lambda = \Lambda(\Omega,\tilde{n},M)$ in $A = D^{n\times n}$ in standard form is given by $\Lambda(\Omega,\tilde{n},\alpha J - M^{tr})$ where $\alpha = 1-m-dm$ and J is the $t{\times}t$-matrix with all entries equal to 1. (In particular $\alpha = 0$ if $D = K$ and $\alpha = -d$ if D is commutative.)*

Proof: The codifferent of Λ is the biggest ideal I of Λ in A with $tr_{A/K}(I) \subseteq R$. By (II.4) (vi) there is a matrix $N = (n_{ij}) \in \mathbf{Z}^{r\times r}$ with $I = \Lambda(\Omega,\tilde{n},N)$. Clearly n_{ii} is the smallest integer satisfying $tr_{D/K}(\mathfrak{P}^{n_{ii}}) \subseteq R$, i.e. $tr_{D/Z}(\mathfrak{P}^{n_{ii}}) \subseteq \bar{\mathfrak{p}}^{-d}$ because $tr_{D/K} = tr_{Z/K} \circ tr_{D/K}$.

But this is equivalent to $tr_{D/Z}(\mathfrak{P}^{n_{ii}+dm}) \subseteq \bar{R}$ since $\bar{\mathfrak{p}}\,\Omega = \mathfrak{P}^m$ (cf. [Rei 75]), which says $n_{ii}+dm = 1-m$, since $\mathfrak{P}^{1-m}$ is the inverse different of Ω over $\bar{R}$. Hence $n_{ii} = 1-m-dm$ for $i=1,\ldots,t$. It follows from the same computation that every twosided ideal $\Lambda(\Omega,n,N')$ of Λ lies in I, whenever $N' = (n'_{ij}) \in \mathbf{Z}^{r\times r}$ satisfies $n'_{ii} \geq n_{ii}$ for $i=1,\ldots,t$. Since $X = \Lambda(\Omega,\tilde{n},\alpha J - M^{tr})$ clearly is a twosided ideal of Λ by (II.4) (vi) it is contained in the inverse different I. It follows from the form of the exponent matrix of X and (II.4) (v) that X - viewed as a left Λ-lattice - is the direct sum of injective indecomposable lattices. Since an injective indecomposable Λ-lattice L has exactly one minimal overlattice in the A-module generated by L (the dual being projective indecomposable), there are exactly t

twosided ideals which are minimal with the property of containing X. The exponent matrices of these ideals are obtained from that of X by decreasing one of the diagonal entries. Hence none of them is contained in I any more, and $I = X$ follows.

q.e.d.

The next topic concerns the interplay between graduated orders Λ and R/p-algebras $\Lambda/\mathfrak{P}\Lambda$. It has been pointed out in [Tar 71] that only finitely many isomorphism types of R/p-algebras can arise in this way if Λ runs through the set of all graduated orders in $A = D^{n\times n}$. Indeed from the discussion of structural invariants one sees: For two graduated orders Λ, Λ' in A with dimension types $\tilde{n}$ and $\tilde{n}'$ and structural invariants $(m_{ijk})_{1\leq i,j,k\leq t}$ resp. $(m'_{ijk})_{1\leq i,j,k\leq t'}$ the R/p-algebras $\Lambda/\mathfrak{P}\Lambda$ and $\Lambda'/\mathfrak{P}\Lambda'$ are isomorphic if and only if the following three conditions are satisfied: (i) $t = t'$, and after a suitable rearrangement of the indices (ii) $\tilde{n} = \tilde{n}'$, (iii) $m_{ijk} = 0$ if and only if $m'_{ijk} = 0$ $(1 \leq i,j,k \leq t)$. (Note, $\Lambda/\mathfrak{P}^2\Lambda$ and $\Lambda'/\mathfrak{P}^2\Lambda'$ are isomorphic as R/p^2-algebras, if and only if in addition to (i)-(iii) also (iv) is satisfied, where (iv) says $m_{ijk} = 1$ iff $m'_{ijk} = 1$ for $1 \leq i,j,k \leq t$. The generalization is obvious and follows from the invariant characterization of the dimension types and the structural invariants of graduated orders. The application of these remarks is simple: If Λ is a graduated order and $\Lambda/\mathfrak{P}\Lambda$ is known then by condition (iii) and (****) one obtains linear equations for the exponent matrix of Λ. A typical example for restricting the possible exponent matrices of Λ by using information about $\Lambda/\mathfrak{P}\Lambda$ is the following lemma, which will be applied in the investigation of blocks with cyclic defect group.

(II.22) Lemma. [Rog 79]. *Let* Λ *be a graduated order in* $A = D^{n\times n}$ *such that every projective indecomposable* $\Lambda/\mathfrak{P}\Lambda$*-module is uniserial (i.e. has a unique composition series), then* $\Lambda \cong \Lambda(\Omega,\tilde{n},\alpha H_t)$ *for a suitable* $\tilde{n} \in \mathbb{N}^{1\times t}$, $\alpha \in \mathbb{N}$, *where*

$$H_t = \begin{pmatrix} 0 & 1 & \cdots & 1 \\ \vdots & \ddots & \ddots & \vdots \\ \vdots & & \ddots & 1 \\ 0 & \cdots & \cdots & 0 \end{pmatrix} \in \mathbb{Z}^{t\times t}$$

has 1's *above the diagonal and* 0's *on and below the diagonal.*

Proof: Let L be a projective indecomposable Λ-lattice and $L_1 = L > L_2 > \ldots > L_{t+1} = L\mathfrak{P}$ the preimage of the unique composition series of $L/L\mathfrak{P}$. Then the $S_i := L_i/L_{i+1}$ for $i = 1,\ldots,t$ form a set of representatives of the simple Λ-(torsion)modules. Define P_i to be the intersection of all Λ-sublattices of L_i which are not contained in L_{i+1} $(i = 1,\ldots,t)$. Since P_i has a unique maximal Λ-sublattice (namely $P_i \cap L_{i+1}$), it is projective, more precisely, it is the projective cover of S_i. If $1 \le j \le i \le t$, then $L/P_j = P_1/P_j$ does not have S_i as composition factor, because otherwise S_i would occur as a composition factor of $L/(P_j+L\mathfrak{P}) = L/L_j$ already (apply the parametrization of irreducible Λ-lattices in (II.4)). Hence $m_{ij1} = 0$ for $1 \le j \le i < t$. The same argument applied to $L = P_k$ yields $m_{ijk} = 0$, if $f(i,k) \ge f(j,k)$ where

$f(i,j) = \begin{cases} i & i \le k \\ i+t & i > k \end{cases}$ for $1 \le i,j \le t$. (Note, the composition factors of $P_k/P_k\mathfrak{P}$ come in the order $S_k, S_{k+1},\ldots,S_t,S_1,\ldots,S_{k-1}$.) Define $m_{ij} = m_{ij1}$ for $1 \le i,j \le t$, and $\alpha = m_{12}$. Then $m_{ij} = 0$ for $1 \le j \le i \le t$. If $1 \le i < k < j \le t$, then

$m_{ijk} = 0 = m_{ij} + m_{jk} - m_{ik} = m_{ij} - m_{ik}$, i.e. the entries in the i-th row above the diagonal are all equal. If $1 \le j < i < k \le t$ then

$m_{ijk} = 0 = m_{ij} + m_{jk} - m_{ik} = m_{jk} - m_{ik}$, i.e. the entries in the k-th row above the diagonal are all equal. Hence all entries above the diagonal are equal to α .

q.e.d.

A second possibility to obtain information about an exponent matrix of a graduated order is to inspect the lattice of Λ-sublattices of an irreducible Λ-lattice.

(II.23) Lemma. Let Λ be a graduated order with structural invariants m_{ijk} , $1 \leq i,j,k \leq t$ and let $S_1,\ldots,S_t$ represent the nonisomorphic simple Λ-(torsion) modules. Up to isomorphism, there are $m_{iji}-1 = m_{jij}-1$ irreducible Λ-lattices L with head $L/\mathrm{Jac}(\Lambda)L$ isomorphic to $S_i \oplus S_j$ $(1 \leq i,j \leq t)$.

Proof: Assume $\Lambda = \Lambda(\Omega,\tilde{n},M)$ to be in standard form and let $V = D^{n\times 1}$ be the usual irreducible A-module. Let $\varepsilon_1,\ldots,\varepsilon_t$ be the standard diagonal idempotents of Λ which yield the central primitive idempotents mod $\mathrm{Jac}(\Lambda)$. Fix $i,j \in \{1,\ldots,t\}$ with $i<j$ and let $\Lambda' = (\varepsilon_i+\varepsilon_j)\Lambda(\varepsilon_i+\varepsilon_j)$. Clearly, $\Lambda' \cong \Lambda(\Omega,(n_i,n_j), \begin{pmatrix} 0 & m_{ij} \\ m_{ji} & 0 \end{pmatrix})$. The parametrization of irreducible lattices in (II.4) shows that Λ' has $m_{ij}+m_{ji}-1$ nonprojective irreducible lattices up to isomorphism. Between these - or more precisely - the nonprojective Λ'-lattices L' in $(\varepsilon_1+\varepsilon_2)V$ and the $L \in \mathfrak{Z}(V)$ with head $S_1 \oplus S_2$ there is a bijection respecting isomorphism: L' is mapped onto $\Lambda L'$; the inverse is given by $L \to (\varepsilon_1+\varepsilon_2)L$. The result follows.

q.e.d.

The extension of this lemma to irreducible Λ-lattices with three simple modules in the head gets slightly more complicated; one obtains a polynomial of degree two for the number of isomorphism classes of irreducible Λ-lattices with head $S_i \oplus S_j \oplus S_k$ $(1 \leq i < j < k \leq t)$, namely $1 - \frac{1}{2}(p_1+p_3) + \frac{1}{4}(p_1{}^2 - p_2)$, where $p_1 = m_{iji} + m_{iki} + m_{jki} + m_{kji}$, $p_2 = (m_{iji} - m_{iki} + m_{jki} - m_{kji})^2$, and $p_3 = m_{iji}{}^2 + m_{iki}{}^2 + (m_{jki} + m_{kji})^2$.

The last topic of this chapter concerns the interplay of graduated orders with general orders. Because graduated orders have a very explicite description one might use them "as a first approximation" to describe a general order. Let Λ and Λ' be two R-orders in the separable (not necessarily simple) K-algebra A. One says Λ' covers Λ, if $\Lambda' \supseteq \Lambda$ and $\mathrm{Jac}(\Lambda') \supseteq \mathrm{Jac}(\Lambda)$, cf. [Rei 75], [Jac 81]. Under these circumstances, one has $\mathrm{Jac}(\Lambda) = \mathrm{Jac}(\Lambda') \cap \Lambda$; hence $\Lambda/\mathrm{Jac}(\Lambda)$ is embedded into $\Lambda'/\mathrm{Jac}(\Lambda')$. Jacobinski calls a minimal hereditary R-order which covers Λ a hereditary hull. It is natural to extend this definition.

<u>*(II.24) Definition.*</u> *Let Λ be an R-order in the separable K-algebra A. A graduated order which is minimal among the graduated orders which cover Λ is called a graduated hull of Λ.*

The existence of graduated hulls follows easily from the existence of hereditary hulls. However, (II.7) and (II.8) do not only yield the existence but also a complete survey over all graduated hulls of an R-order Λ. Note, a graduated order in A contains the primitive central idempotents ε_s, $s = 1,\ldots,h$ of A. Therefore a graduated hull of Λ is necessarily a graduated hull of $\bigoplus_{s=1}^{h} \varepsilon_s\Lambda$ and vice versa. This reduces the discussion of graduated hulls of general orders to those of orders Λ in simple algebras. $A = D^{n\times n}$. Let V be an irreducible A-module and $\mathfrak{Z}(V)$ the $(\cap,+)$-lattice of all Λ-Ω-lattices $\neq 0$ in V. Clearly $\mathfrak{Z}(V)$ is an admissible set of lattices in the sense defined before (II.7). Call a $(\cap,+)$-sublattice $\mathfrak{Z}$ of $\mathfrak{Z}(V)$ an admissible distributive sublattice of $\mathfrak{Z}(V)$ if it satisfies the three conditions of an admissible set of Ω-lattices and if it is distributive as a $(\cap,+)$-lattice. The order

$$\mathfrak{O}(\mathfrak{Z}) = \{x \in A \mid xL \subseteq L \quad \text{for all} \quad L \in \mathfrak{Z}\}$$

is a graduated overorder of Λ by (II.8) for each admissible distributive sublattice $\mathfrak{Z}$ of $\mathfrak{Z}(V)$. On the other hand, for each graduated order Γ in A containing Λ, the set $\mathfrak{L}(\Gamma)$ of all Γ-lattices $\neq 0$ in V is an admissible distributive sublattice of $\mathfrak{Z}(V)$ by (II.4). This almost proves the first part of the final result of this chapter.

<u>*(II.25) Proposition.*</u> *Let Λ be an R-order in $A = D^{n\times n}$, V an irreducible A-module, and $\mathfrak{Z}(V)$, $\mathfrak{O}$ and $\mathfrak{L}$ as just defined. Then*

(i) $\mathfrak{L}$ and $\mathfrak{O}$ define a Galois correspondence between the graduated R-orders Γ in A containing Λ and the admissible distributive sublattices of $\mathfrak{Z}(V)$.

(ii) The maximal admissible distributive sublattices of $\mathfrak{Z}(V)$ correspond to the graduated hulls of Λ. In particular, Λ has exactly one graduated hull, if and only if $\mathfrak{Z}(V)$ is distributive.

Proof: (i) Clearly $\mathfrak{L}$ and $\mathfrak{O}$ reverse inclusions and $\mathfrak{L}(\mathfrak{O}(\mathfrak{Z})) = \mathfrak{Z}$ and $\mathfrak{O}(\mathfrak{L}(\Gamma)) = \Gamma$ for all admissible distributive sublattices $\mathfrak{Z}$ of $\mathfrak{Z}(V)$ and all graduated overorders Γ of Λ.

(ii) Let $\mathfrak{Z}$ be a maximal admissible distributive sublattice of $\mathfrak{Z}(V)$ and $\Gamma = \mathfrak{O}(\mathfrak{Z})$. By (i) Γ is a minimal graduated overorder of Λ. To prove that Γ is a graduated hull of Λ it suffices to prove $\mathrm{Jac}(\Gamma) \supseteq \mathrm{Jac}(\Lambda)$. Let $L_i \in \mathfrak{Z}$ with $L_0 = L > L_1 > \ldots > L_t = L\mathfrak{P}$ be a Γ-composition series of $L/L\mathfrak{P}$. Suppose there exists an i, $0 \leq i < t$, and an $L' \in \mathfrak{Z}(V)$ with $L_{i+1} \subsetneqq L' \subsetneqq L_i$. This contradicts the maximality of $\mathfrak{Z}$, namely $\mathfrak{Z}$ together with the $L'\mathfrak{p}^{\alpha}$ $(\alpha \in \mathbb{Z})$ would still generate a distributive admissible sublattice of $\mathfrak{Z}(V)$. Now one easily concludes that L_i/L_{i+1} as a Λ-module splits into a direct sum of isomorphic irreducible Λ-modules for $i = 0,\dots,t-1$. Hence

$$\mathrm{Jac}(\Gamma) = \{x \in \Gamma \mid x\, L_i \subseteq L_{i+1} \text{ for } i = 0,\dots,t-1\} \supseteq$$
$$\{x \in \Lambda \mid x\, L_i \subseteq L_{i+1} \text{ for } i = 0,\dots,t-1\} = \mathrm{Jac}(\Lambda) .$$

q.e.d.

III. The conductor formula for graduated hulls of selfdual orders

In this chapter R is a complete local Dedekind domain with quotient field K and maximal ideal $\mathfrak{p}$. The residue class field $F = R/\mathfrak{p}$ is assumed to be finite. $A = \bigoplus_{s=1}^{h} A_s$ is a separable K-algebra with minimal twosided ideals $A_s \cong D_s^{l_s \times l_s}$ where D_s is a division algebra over K with center Z_s and dimension m_s^2 over Z_s. For $1 \leq s \leq h$ the relative reduced trace of A_s (i.e. the composition of the reduced trace tr_{A_s/Z_s} and $tr_{Z_s/K}$) is denoted by tr_s. For a unit $u = u_1 + \ldots + u_h$ in the center $Z(A)$ of A $(u_s \in Z_s, u_s \neq 0)$ denote by T_u the generalized trace map of A into K defined by $T_u(a) = \sum_{s=1}^{h} tr_s(u_s a_s)$, where $a = a_1 + \ldots + a_h$ with $a_s \in A_s$. Each T_u induces a nondegenerate symmetric bilinear form $\Phi_u : A \times A \to K : (a,b) \to T_u(ab)$ satisfying $\Phi_u(ab,c) = \Phi_u(a,bc)$ for all $a,b,c \in A$.

For any subset M of A, let M^* be the dual R-module of M with respect to T_u defined by $M^* = \{a \in A \mid T_u(am) \in R \text{ for all } m \in M\}$. Note, if M is a full R-lattice in A (i.e. $\operatorname{rank}_R M = \dim_K A$), the same holds for M^*. Moreover, $M^{**} = M$ in this case.

<u>*(III.1) Definition.*</u> *An R-order Λ in A is called selfdual, if there exists a unit $u \in Z(A)$ such that the dual Λ^* of Λ with respect to T_u coincides with Λ.*

Note, if Λ is selfdual, the same holds for the block ideals of Λ. The major examples of selfdual orders are group rings, cf. [Jac 81]. The "twisted" group rings include these examples.

<u>*(III.2) Example.*</u> *Let G be a finite group and α a 2-cocycle of G with values in the unit group $U(R)$ of R. (G acts trivially on*

$U(R)$.) Let $A = K(G,\alpha)$ be a $|G|$-dimensional K-algebra with K-basis $v(g)$, $g \in G$, and multiplication rule $v(g)v(h) = \alpha(g,h)v(gh)$, where K is a finite extension field of the p-adics $\mathbb{Q}_p$. Then $K(G,\alpha)$ is a separable K-algebra and $\Lambda = R(G,\alpha) = \bigoplus_{g\in G} Rv(g)$ is a selfdual R-order in A (with respect to $\frac{1}{|G|}Tr$, where Tr is the regular trace of A.)

Proof: Assume without loss of generality $\alpha(g,1) = \alpha(1,g) = 1$ for all $g \in G$. Then $Tr(v(g)) = |G|$ for $g = 1$ and zero for $g \in G$, $g \neq 1$. Hence, the discriminant $\det(Tr(v(g)v(h)))_{g,h\in G} = \pm |G|^{|G|} \cdot \prod_{g\in G} \alpha(g,g^{-1})$ with respect to the trace Tr of the regular representation is not zero, since $\alpha(g,g^{-1}) \in U(R)$. Hence, A is separable. Clearly $\Lambda \subseteq \Lambda^*$ where Λ^* is the dual with respect to $\frac{1}{|G|}Tr$. Conversely, let $x = \sum_{g\in G} a_g v(g) \in \Lambda^*$. Then $\alpha(g^{-1},g)a_g \in R$; hence $a_g \in R$ for all $g \in G$, since $\alpha(g^{-1},g)$ is a unit in R. This proves $\Lambda^* = \Lambda$. Note, $\frac{1}{|G|}Tr = T_u$ with $u = \sum_{s=1}^{h} \frac{n_s}{|G|}\varepsilon_s$, where $n_s = m_s l_s$ is the $\tilde{K}$-dimension of an (absolutely) irreducible $\tilde{K} \otimes_K A_s$-module for some splitting field $\tilde{K}$ of A_s, and ε_s is the central primitive idempotent of A with $\varepsilon_s A = A_s$, cf. e.g. [Rei 75], [Jac 81].

q.e.d.

<u>*(III.3) Lemma.*</u> *Let Λ be a selfdual R-order in A, let K' be a subfield of K with $[K : K'] < \infty$ such that K is separable over K', and let $R' = K' \cap R$. Then Λ is also a selfdual R'-order.*

Proof: Let $u \in Z(A)$ such that the dual Λ^* with respect to T_u is equal to Λ. Let $u' \in K$ be a generator of the inverse different of R over R' (with respect to $tr_{K/K'}$). Define $T'_{u'u} : A \to K'$ by $T'_{u'u}(a) = tr_{K/K'}(u'T_u(a))$ for all $a \in A$. Then the dual of Λ as R'-lattice with respect to $T'_{u'u}$ is also equal to Λ.

q.e.d.

<u>*(III.4) Corollary.*</u> *Let H be a finite group with nontrivial center $Z(H)$ (not necessarily contained in the derived subgroup H'), let K be a finite unramified extension field of $\mathbb{Q}_p$, p a rational prime. For every primitive idempotent ε in the group algebra $KZ(H)$ $(\subseteq KH)$ of the center of H the R-order εRH in εKH is selfdual.*

Proof: $\tilde{K} = \varepsilon KZ(H)$ is an extension field of K which arises by adjoining certain (not necessarily primitive) $|H|$-th roots of unity to K . The R-order $\tilde{R} = \varepsilon RZ(H)$ is maximal in $\tilde{K}$, since K is unramified over $\mathbb{Q}_p$. Since $\tilde{R} \subseteq \varepsilon RH$, one can view εRH as an $\tilde{R}$-order, and as such it is of the type described in (III.2), namely $\varepsilon RH = R(G,\alpha)$ for $G = H/Z(H)$ and α suitable. Hence, by (III.2) and (III.3) εRH is also a selfdual R-order.

q.e.d.

Though it is not needed in this paper, the following criterion might be useful.

<u>*(III.5) Proposition.*</u> *Let Λ be an R-order in A , $\tilde{K}/K$ a finite field extension, and let $\tilde{R}$ be the integral closure of R in $\tilde{K}$. Then Λ is a selfdual R-order in A if and only if $\tilde{\Lambda} = \tilde{R} \otimes_R \Lambda$ is a selfdual $\tilde{R}$-order in $\tilde{A} = \tilde{K} \otimes_K A$.*

Proof: If Λ is selfdual with respect to T_u , u a unit in $Z(A)$, then $\tilde{\Lambda}$ is selfdual with respect to the generalized trace map $T_{1 \otimes u}$ of $\tilde{A}$. Conversely, assume $\tilde{\Lambda}$ is selfdual with respect to $T_{\tilde{u}}$ for a suitable unit $\tilde{u} \in Z(\tilde{A})$. Denote by Γ resp. $\tilde{\Gamma}$ the maximal order of $Z(A)$ resp. $Z(\tilde{A})$. Let $u \in Z(A)$ with $u\Gamma = \tilde{u}\tilde{\Gamma} \cap A$. (A is identified with the subset $1 \otimes_K A$ of $\tilde{A}$.) Then the generalized trace map T_u of A has the property that the dual Λ^* of Λ in A with respect to T_u is equal to Λ . To prove this let $\lambda_1,\ldots,\lambda_r$ be an R-basis of Λ , and $\lambda_1^*,\ldots,\lambda_r^* \in A$ be defined by $T_u(\lambda_i\lambda_j^*) = \delta_{ij}$ (Konecker δ) for $1 \leq i,j \leq r$. Similarly define $\lambda_1',\ldots,\lambda_r' \in \tilde{A}$ by $T_{\tilde{u}}(\lambda_i\lambda_j') = \delta_{ij}$ $(1 \leq i,j \leq r)$. Then $\lambda_j^* = \tilde{u}u^{-1}\lambda_j'$ for $j = 1,\ldots,r$. Hence,

$$\Lambda^* = \bigoplus_{i=1}^{r} R\lambda_i^* = \bigoplus_{i=1}^{r} \tilde{R}\lambda_i^* \cap A = (\tilde{u}u^{-1} \bigoplus_{i=1}^{r} \tilde{R}\lambda_i') \cap A = \tilde{u}u^{-1}\tilde{\Lambda} \cap A = \bigoplus_{i=1}^{r} (\tilde{u}u^{-1}\tilde{R}\lambda_i \cap K\lambda_i) =$$

$$= \bigoplus_{i=1}^{r} R\lambda_i = \Lambda \ .$$

q.e.d.

The following result will be applied later on and, together with the conductor formula (III.8), it might justify the introduction of selfdual orders.

(III.6) Proposition. *Let* Λ *be a selfdual order in* A *and* $\varepsilon \in \Lambda$ *an idempotent of* Λ *. Then* $\varepsilon\Lambda\varepsilon$ *is a selfdual order in* $\varepsilon A\varepsilon$ *.*

Proof: Let u be a unit in $Z(A)$ such that the dual Λ^* of Λ with respect to T_u is equal to Λ . Then $\varepsilon u \in Z(\varepsilon A\varepsilon)$, $T_{\varepsilon u} : \varepsilon A\varepsilon \to K : a \to T_u(a)$ is a generalized trace map of $\varepsilon A\varepsilon$. Moreover, the dual of $\varepsilon\Lambda\varepsilon$ with respect to $T_{\varepsilon u}$ is equal to $\varepsilon\Lambda\varepsilon$.

q.e.d.

As a consequence of (III.6) one sees that selfduality is compatible with Morita equivalence: If Λ_1 and Λ_2 are Morita equivalent orders such that Λ_1 is selfdual, then Λ_2 is selfdual.

Now, let Λ be a selfdual R-order in A . For the investigation of Λ the following simple observation is basic: If $M \subseteq A$ is a right resp. left Λ-module, then M^* is a left resp. right Λ-module. If

Γ is an R-order in A which contains Λ, the biggest right resp. left Γ-ideal contained in Λ is called the right resp. left conductor of Γ in Λ; cf. [Jac 81]. The following result is due to Jacobinski in case of group rings, cf. [Jac 81]. In the situation of selfdual orders the proof is the same and can be omitted.

<u>*(III.7) Proposition.*</u> *Let Λ be a selfdual R-order in A and Γ an R-order in A containing Λ. The right and the left conductor of Γ in Λ coincide and are given by Γ^*. (Γ^* is called the conductor of Γ in Λ in this situation.)*

(III.7) and (II.21) allow to give an explicite formula for the conductor of Γ in Λ, in case Γ is a graduated R-order. Thus, the next result generalizes Jacobinski's conductor formula for hereditary overorders of group rings, cf. [Jac 66], [Jac 81].

<u>*(III.8) Theorem.*</u> *(Conductor formula). Let Λ be an R-order in A which is selfdual with respect to T_u ($u = u_1+\ldots+u_h$ with $u_s \in Z_s$, $u_s \neq 0$ for $s = 1,\ldots,h$). Let $\Gamma = \bigoplus_{s=1}^{h} \Gamma_s$ be a graduated order in A containing Λ (with $\Gamma_s \subseteq A_s$). The conductor of Γ in Λ is given by $\Gamma^* = \bigoplus_{s=1}^{h} u_s^{-1}\mathfrak{D}_s^{-1}\mathfrak{D}_{A_s/Z_s}(\Gamma_s) = \bigoplus_{s=1}^{h} u_s^{-1}\mathfrak{D}_{A_s/K}(\Gamma_s)$*

where $\mathfrak{D}_s$ is the different of the integral closure R_s of R in Z_s over R, and $\mathfrak{D}_{A_s/Z_s}(\Gamma_s)$ resp. $\mathfrak{D}_{A_s/K}(\Gamma_s)$ is the codifferent of Γ_s in A_s over Z_s resp. K (cf. (II.21)).

If Γ_s is isomorphic to the graduated order $\Lambda(\Omega_s,\tilde{n}_s,M^{(s)})$ in standard form, where Ω_s is the maximal R-order in D_s, $\tilde{n}_s \in \mathbf{N}^{1\times t_s}$, and $M^{(s)} \in \mathbf{Z}_{\geq 0}^{t_s\times t_s}$ for suitable $t_s \in \mathbf{N}$, then - by the same isomorphism - the s-th component of the conductor Γ^* corresponds to

$\Lambda(\Omega_s, \tilde{n}_s, \varkappa_s J_{t_s} - M^{(s)^{tr}})$, where J_{t_s} is the $t_s \times t_s$-matrix with all entries equal to 1 and where $\varkappa_s$ is determined as follows $(1 \leq s \leq h)$:

$\varkappa_s = 1 + m_s(\mu_s - d_s - 1)$ with $m_s^2 = \dim_{Z_s} D_s$, $\mathfrak{p}_s^{d_s} = \mathfrak{D}_s$ and $\mathfrak{p}_s^{\mu_s} = u_s^{-1} R_s$ with $\mathfrak{p}_s = Jac(R_s)$.

Proof: By (III.7) the conductor of Γ in Λ is given by Γ^*. One has $x = \sum_{s=1}^{h} x_s \in \Gamma^*$ $(x_s \in A_s, s = 1,\ldots,h)$ iff $T_u(xy) = \sum_{s=1}^{h} tr_s(u_s x_s y) \in R$ for all $y \in \Gamma$. This holds iff $tr_{Z_s/K}(u_s tr_{A_s/Z_s}(x_s y_s)) \in R$ for all $y_s \in \Gamma_s$, which is equivalent to $tr_{A_s/Z_s}(x_s y_s) \in u_s^{-1}\mathfrak{D}_s^{-1}$ for all $y_s \in \Gamma_s$, i.e. $x_s \in u_s^{-1}\mathfrak{D}_s^{-1}\mathfrak{D}_{A_s/Z_s}(\Gamma_s) = u_s^{-1}\mathfrak{D}_{A_s/K}(\Gamma_s)$. This proves the first part of the theorem, the rest follows from (II.21).

q.e.d.

A generalization of the conductor formula to graduable rather than graduated overorders can easily be obtained from this conductor formula, (II.19) and (II.20).

As a consequence of (III.8) one gets the inequalities $m_{iji}^{(s)} = m_{ij}^{(s)} + m_{ji}^{(s)} \leq \varkappa_s$ which restrict the possibilities for $M^{(s)}$.

Note, in the examples discussed, u_s, and hence the constant μ_s in the last theorem, is easily computed. Namely, in the case of (III.2), which includes ordinary group rings RG, one has $u_s = \frac{1_s m_s}{|G|}\varepsilon_s$, where ε_s is the central idempotent of A in A_s. Note, $1_s m_s$ is the degree of an absolutely irreducible $\tilde{K} \otimes_K A_s$-representation for some splitting field $\tilde{K}$ of A_s. In the case of Corollary (III.4) the proof of (III.3) shows $u_s = \frac{l_s m_s}{|G|} x\varepsilon_s$, where $G = H/Z(H)$ and where x generates the inverse different of $\tilde{R} = \varepsilon RZ(H)$ over R .

Coming back to the general case of a selfdual R-order Λ , it is worthwhile to investigate the Λ-Λ-bimodules Γ/Λ and Λ/Γ^* , where Γ is an R-order containing Λ . (Of course, a graduated hull of Λ is a good candidate for Γ .)

Let $\bigoplus_{i=1}^{r} \bar{\Lambda}_i$ be the decomposition of $\Lambda/\mathrm{Jac}(\Lambda)$ into minimal twosided ideals, and let M_{ij} , $1 \leq i,j \leq r$, be representatives for the simple Λ-Λ-(torsion)modules such that $\bar{\Lambda}_i M_{ij} \bar{\Lambda}_j = M_{ij}$.

<u>(III.9) Remark.</u> *The multiplicity of* M_{ij} *in a* Λ-Λ-*composition series of* Γ/Λ *is the same as the multiplicity of* M_{ji} *in* Λ/Γ^* .

Proof: Let $\Gamma = X_1 > X_2 > \ldots > X_l = \Lambda$ be a Λ-Λ-composition series of Γ/Λ . Then $\Lambda = \Lambda^* = X_l^* > X_{l-1}^* > \ldots > X_1^* = \Gamma^*$ is a composition series for Λ/Γ^* . Moreover, if X_t/X_{t+1} is isomorphic to M_{ij} , then X_{t+1}^*/X_t^* is isomorphic to M_{ji} for $0 \leq t \leq l$ and $1 \leq i,j \leq r$.

q.e.d.

A particularly natural choice for an overorder Γ of Λ is $\Gamma = \bigoplus_{s=1}^{h} \varepsilon_s \Lambda$, where the $\varepsilon_s \in A_s$ $(1 \leq s \leq h)$ are the central primitive idempotents of A . Clearly, in this case the conductor of Γ in Λ is given by $\bigoplus_{s=1}^{h} (\Lambda \cap \varepsilon_s \Lambda)$, which will be of importance in the next chapter. The next topic of this chapter is devoted to the question when $\Gamma = \bigoplus_{s=1}^{h} \varepsilon_s \Lambda$ is a graduated order. Clearly in this situation Γ is the unique graduated hull of Λ . For the following it need not be assumed that Λ is selfdual. It is convenient to define decomposition numbers in a slightly different way as usual, cf. [Jac 81]. Some additional notation is needed.

Let Λ be an R-order in A and $\Theta = \bigoplus_{s=1}^{h} \Theta_s$ with $\Theta_s \subseteq A_s$ $(1 \leq s \leq h)$ a maximal R-order in A containing Λ. Let $S_1, \ldots, S_r$ be representatives of the isomorphism classes of the simple Λ-(torsion)-modules, and $\tilde{S}_1, \ldots, \tilde{S}_h$ with $\Theta_s \tilde{S}_s = \tilde{S}_s$ $(1 \leq s \leq h)$ representatives of the simple Θ-modules. Following JACOBINSKI, the decomposition numbers of Λ are defined as follows, cf. [Jac 81].

<u>(III.10) Definition.</u> *The multiplicity* d_{si} *of* S_i *in the restriction* $\tilde{S}_{s|\Lambda}$ *of* $\tilde{S}_s$ *to* Λ *is called a decomposition number (of* Λ*) for* $1 \leq s \leq h$, $1 \leq i \leq r$.

An alternative definition (not used in the sequel) of the decomposition numbers can be given by using a maximal separable suborder Σ of Λ as introduced in II.c , cf. [Jac 81b]. Namely, since the components Σ_i of Σ are in 1-1-correspondence with the S_i , the decomposition number d_{si} can be viewed as the multiplicity of an irreducible Σ-lattice L_i with $\Sigma_i L_i = L_i$ as a component in the restriction of the irreducible Θ-lattice $\tilde{L}_s$ with $\Theta_s \tilde{L}_s = \tilde{L}_s$ to Σ .

The usual argument shows that the d_{si} are independent of the choice of the maximal order Θ . Note, multiplying d_{si} by the ramification index of the maximal order Ω_s of D_s over R (which is equal to m_s multiplied by the ramification index of Z_s over K) yields the decomposition numbers as they are usually defined, cf. e.g. [Ser 77], [Gre 74]. Denote the projective cover of S_i by P_i $(1 \leq i \leq r)$ and let $V_1, \ldots, V_h$ be irreducible A-modules with $A_s V_s = V_s$ $(1 \leq s \leq h)$. If d'_{si} is the multiplicity of V_s in the A-module $KP_i := K \otimes_R P_i$, then the famous Brauer reciprocity reads as follows:

(III.11) Theorem. *(Brauer reciprocity)* $f_s d'_{si} = d_{si} \dim_F \mathrm{End}_\Lambda(S_i)$ *for* $1 \leq i \leq r$, $1 \leq s \leq h$, *where* $f_s = \dim_F \Omega_s/\mathfrak{P}_s = \dim_F \mathrm{End}_\Theta(\tilde{S}_s)$ *with* Ω_s *as maximal R-order in* D_s *and* $\mathfrak{P}_s = \mathrm{Jac}(\Omega_s)$.

Proof: The usual equation of Brauer reciprocity, cf. e.g. [Gre 74], is divided by the ramification index of Ω_s over R and reinterpreted as above.

q.e.d.

The desired criterion for $\bigoplus_{s=1}^{h} \varepsilon_s\Lambda$ to be a graduated or at least a graduable order can be obtained from the next result.

(III.12) Proposition. *Let* ε_s *be the central primitive idempotent of* A *lying in* A_s $(1 \leq s \leq h)$.

(i) $\varepsilon_s\Lambda$ *is a graduable order if and only if* $d_{si} \leq 1$ *for* $i = 1,\ldots,r$ *and* $Z(\varepsilon_s\Lambda)$ *is the maximal order in the center* Z_s *of* A_s .

(ii) Let $\varepsilon_s\Lambda$ *be graduable and* Ω_s *commutative.* $\varepsilon_s\Lambda$ *is a graduated R-order iff* $\dim_F \mathrm{End}_\Lambda(S_i) = f_s$ *for* $i = 1,\ldots,r$ *whenever* $d_{si} = 1$, *where* f_s *is the inertial degree defined in (III.11).*

Proof: (i) This follows from (II.16) (equivalence of (i') and (v')).
(ii) This follows from (II.8) and the discussion following (II.8).

q.e.d.

Note, in (III.12) (i) $d_{si} \leq 1$ for $i = 1,\ldots,r$ implies already that $Z(\varepsilon_s\Lambda)$ is the maximal order in Z_s if Ω_s is unramified over R . This follows easily by inspecting the semisimple $R/\mathfrak{p}$-subalgebras of

$(\Omega_s/\mathrm{Jac}(\Omega_s))^{l_s \times l_s}$. Often, but not always, the multiplication by ε_s maps the center of Λ onto the center of $\varepsilon_s\Lambda$. In case of group rings one therefore can often decide positively that $Z(\varepsilon_s\Lambda)$ is a maximal order by looking at the central characters of the group. Once the decomposition numbers are known, it is easy to compute $\dim_F \mathrm{End}_\Lambda(S_i)$ by looking at the values of the Brauer characters in case Λ is a group ring. However, it might not be clear whether or not Ω_s is commutative, more precisely what the Schur index of D_s is. The following remark might be helpful.

(III.13) Remark. *Assume the decomposition number d_{si} of Λ , for one pair of indices s,i , $1 \le s \le h$, $1 \le i \le r$, is equal to 1 . Assume as in (III.12) (i) that the center of $\varepsilon_s\Lambda$ is equal to the maximal R-order R_s of Z_s and let $\tilde{\varepsilon}_i$ be an idempotent of Λ with $\tilde{\varepsilon}_i S_i = S_i$ and $\tilde{\varepsilon}_i S_j = 0$ for $j \neq i$, $1 \le j \le r$.*

(i) If $\Gamma_{s,i} = \tilde{\varepsilon}_i \varepsilon_s \Lambda \tilde{\varepsilon}_i$ is a maximal R-order, then the index m_s of D_s $(= \sqrt{\dim_{Z_s} D_s})$ is given by the quotient $\dfrac{\dim_F \mathrm{End}_\Lambda(S_i)}{\dim_F R_s/\mathrm{Jac}(R_s)}$.

(ii) $\Gamma_{s,i}$ as defined in (i) is a maximal R-order iff for some unramified extension field $\tilde{K}$ of K with maximal order $\tilde{R}$ the $\tilde{R}$-order $\tilde{\Gamma}_{s,i} = \tilde{R} \otimes_R \Gamma_{s,i}$ is a hereditary order.

Proof: (i) This is an immediate consequence of the classification of maximal R-orders, cf. e.g. [Rei 75], (note, $\Gamma_{s,i}$ is an R_s-order and the quotient of the dimensions is equal to $\dim_{R_s/\mathrm{Jac}(R_s)} \mathrm{End}_{\Gamma_{s,i}}(S_i)$.)

(ii) Maximal orders are hereditary. By a theorem of Auslander and Goldman (cf. [Rei 75], Thm. 39.1) an order Γ is hereditary iff $\mathrm{Jac}(\Gamma)$ is projective.

Since $\tilde{R} \otimes_R \mathrm{Jac}(\Gamma_{s,i}) = \mathrm{Jac}(\tilde{\Gamma}_{s,i})$ and since local hereditary orders are maximal, (ii) follows.

q.e.d.

The way this remark can be applied e.g. for group rings RG with sufficiently many decomposition numbers equal to 1 is demonstrated in Chapter VII: One first computes the graduated hull of $\tilde{R}G$ for "sufficiently big" $\tilde{R}$ unramified over R and then comes back to R . Though the results of Chapter II.b on graduable orders, in particular (II.20), yield the more general and more precise results in this context, it often suffices to apply the more elementary Remark (III.13).

The last of the ingredients in the constant $\varkappa_s = 1 - m_s(\mu_s - d_s - 1)$ in the conductor formula (III.8) to be commented upon is d_s or the different $\mathfrak{p}_s^{d_s}$ of R_s over R , ($\mathfrak{p}_s = \mathrm{Jac}(R_s)$). If the characteristic of K is zero, K is a finite extension of the p-adic field $\mathbb{Q}_p$. In the case of group rings the interesting case occurs when K is unramified over $\mathbb{Q}_p$. Clearly, in this case Z_s is contained in an extension field $K[\zeta]$ of K , where ζ is suitable (e.g. $|G|$-th) root of unity. Since the different of an unramified extension is trivial, d_s can be obtained from the different $\mathfrak{D}(K_s/\mathbb{Q}_p)$ of the unique subfield K_s of $\mathbb{Q}_p[\zeta_\alpha]$ with $Z_s = K_s K_s'$, where ζ_α is a primitive p^α-th root of unity and K_s' is the inertia field of Z_s with respect to $\mathbb{Q}_p$: One has $\mathfrak{D}(Z_s/K) = \mathfrak{D}(K_s/\mathbb{Q}_p)R_s$. Hence Corollary (III.15) below, which certainly can also be found in the number theoretic literature, cf. e.g. [Was 82], Thm. 3.11, solves the determination of the d_s completely for group rings (Note, Z_s can be read off from the character table.).

(III.14) Lemma. Let p be a prime number, R the ring of p-adic integers, and C_{p^α} the cyclic group of order p^α .

(i) $RC_{p^\alpha} \cong \{(a,b) \in RC_{p^{\alpha-1}} \oplus R_\alpha \mid \varphi_\alpha(a) = \psi_\alpha(b)\}$.

Where $R_\alpha = R[\zeta_\alpha]$ the maximal R-order in $\mathbb{Q}_p[\zeta_\alpha]$, ζ_α a p^α-th root of unity, and where $\varphi_\alpha : RC_{p^{\alpha-1}} \to FC_{p^{\alpha-1}}$ and $\psi_\alpha : R_\alpha \to FC_{p^{\alpha-1}}$

are the obvious epimorphisms. $(\alpha \geq 1)$.

(ii) *Let* p *be odd,* d *a divisor of* $p-1$, *and denote by* $G_{d,\alpha}$ *the semidirect product of* C_{p^α} *with* C_d , *where* C_d *acts faithfully on* C_{p^α} . *The cyclic group* C_d *acts faithfully on the field* $\mathbb{Q}_p[\zeta_\alpha]$ *with fixed subfield* $K_{d,\alpha}$. *Denote the maximal R-order of* $K_{d,\alpha}$ *by* $R_{d,\alpha}$. *Then there are epimorphisms* $\varphi_{\alpha,d} : RG_{d,\alpha-1} \to FG_{d,\alpha-1}$ *and* $\psi_{\alpha,d} : \Lambda(R_{d,\alpha}, \underbrace{1,\dots,1}_{d}, H_d) \longrightarrow$

$$\to FG_{d,\alpha-1} \quad (H_d = \begin{pmatrix} 0 & 1 & \cdots & 1 \\ \vdots & \ddots & \ddots & \vdots \\ & & \ddots & 1 \\ 0 & \cdots & \cdots & 0 \end{pmatrix} \in \mathbb{Z}^{d\times d}), \text{ with } RG_{d,\alpha} \cong$$

$\{(a,b) \in RG_{d,\alpha-1} \oplus \Lambda(R_{d,\alpha}, 1,\dots,1, H_d) \mid \varphi_{\alpha,d}(a) = \psi_{\alpha,d}(b)\}$. $(\alpha \geq 1)$.

(iii) *Let* $p=2$, σ *one of the three automorphisms of order* 2 *of* $\mathbb{Q}_p[\zeta_\alpha]$, $(\sigma\colon \zeta_\alpha \to \zeta_\alpha^{-1}$, *or* $\sigma\colon \zeta_\alpha \to \zeta_\alpha^{1+2^{\alpha-1}}$, *or* $\sigma\colon \zeta_\alpha \to \zeta_\alpha^{-1+2^{\alpha-1}})$, $K_{\sigma,\alpha}$ *the fixed subfield of* $\mathbb{Q}_p[\zeta_\alpha]$ *under* σ, $R_{\sigma,\alpha}$ *the maximal R-order in* $K_{\sigma,\alpha}$, *and* $G_{\sigma,\alpha}$ *the semi direct product of* $\langle\zeta_\alpha\rangle \cong C_{p^\alpha}$ *by* $\langle\sigma\rangle$ *(with the above action of* σ*). Then there is an R-order* $X_{\sigma,\alpha}$ *with* $\Lambda(R_{\sigma,\alpha}, 1, 1, \begin{pmatrix} 0 & 1 \\ 0 & 0 \end{pmatrix})$ *as unique graduated hull and center* $R_{\sigma,\alpha}$ *such that there are epimorphisms* $\varphi_{\sigma,\alpha} : RG_{\sigma,\alpha-1} \to FG_{\sigma,\alpha-1}$ *and* $\psi_{\sigma,\alpha} : X_{\sigma,\alpha} \to FG_{\sigma,\alpha-1}$ *with* $RG_{\sigma,\alpha} \cong \{(a,b) \in RG_{\sigma,\alpha-1} \oplus X_{\sigma,\alpha} \mid \varphi_{\sigma,\alpha}(a) = \psi_{\sigma,\alpha-1}(b)\}$. $(\alpha \geq 2)$.

<u>Proof:</u> (i) Let $C_{p^\alpha} = \langle x_\alpha \rangle$; then $x_\alpha \to x_{\alpha-1}$ and $x_\alpha \to \zeta_\alpha$ induce the desired epimorphisms. Hence to prove that RC_{p^α} and the described amalgam of $RC_{p^{\alpha-1}}$ and R_α are isomorphic, it suffices to show that $RC_{p^{\alpha-1}}$ and R_α have no common epimorphic images bigger than $FC_{p^{\alpha-1}}$. But for $p \neq 2$, $FC_{p^{\alpha-1}}$ is the biggest factor module of $RC_{p^{\alpha-1}}$ annihilated by $\mathfrak{p} = pR$ and R_α has a uniserial lattice of

ideals with $\mathfrak{p}R$ properly contained in kernel of ψ_α (since $p^{\alpha-1} < p^\alpha - p^{\alpha-1}$.). In case $p=2$ one has to observe that C_{p^α} acts already faithfully on R_α / I for $I \subsetneqq R_\alpha$, whereas it does not on $RC_{p^{\alpha-1}}$.

(ii) Clearly the conditions of (III.12) are satisfied and the irreducible $RG_{d,\alpha}$-lattices are uniserial, which shows that $RG_{d,\alpha}$ maps onto the hereditary order $\Lambda(R_{d,\alpha},1,\ldots,1,H_d)$. That the amalgamation is of the described type follows by the same argument as in (i) (applied to the d projective indecomposable $RG_{d,\alpha}$-lattices).

(iii) Again the irreducible $RG_{\sigma,\alpha}$-lattices are uniserial (as one sees by restricting the operation to C_{p^α}) and the center of $RG_{\sigma,\alpha}$ maps onto $R_{\sigma,\alpha}$. This proves that $\Lambda(R_{\sigma,\alpha},1,1,\begin{pmatrix}0&1\\0&0\end{pmatrix})$ is the unique graduated hull of $X_{\sigma,\alpha}$, which is now defined as the obvious epimorphic image of $RG_{\sigma,\alpha}$, and that $R_{\sigma,\alpha}$ is the center of $X_{\sigma,\alpha}$. Note $G_{\sigma,\alpha-1} \cong G_{\sigma,\alpha}/Z(G_{\sigma,\alpha})$ and $|Z(G_{\sigma,\alpha})| = 2$. Therefore $Z(G_{\sigma,\alpha})$ is in the kernel of the action of $G_{\sigma,\alpha}$ on $X_{\sigma,\alpha}/2X_{\sigma,\alpha}$; however, $G_{\sigma,\alpha}$ acts faithfully on bigger factor modules of $X_{\sigma,\alpha}$.

Hence $FG_{\sigma,\alpha-1} \cong X_{\sigma,\alpha}/2X_{\sigma,\alpha}$ is the biggest common $RG_{\sigma,\alpha-1}$-factor module of $RG_{\sigma,\alpha-1}$ and $X_{\sigma,\alpha}$. (Note, $X_{\sigma,\alpha}$ can be described explicitely as a subring of $\Lambda(R_{\sigma,\alpha},1,1,\begin{pmatrix}0&1\\0&0\end{pmatrix})$, if one compares which are the biggest isomorphic factor modules of the two irreducible $X_{\sigma,\alpha}$-lattices; the answer depends on σ, cf. comments following the proof of (III.15).)

q.e.d.

<u>(III.15) Corollary:</u> Let $\tilde{K}$ be any subfield of $\mathbb{Q}_p[\zeta_\alpha]$ containing $\mathbb{Q}_p$, where ζ_α is a primitive p^α-th root of unity. The different $\tilde{\mathfrak{p}}^d$ of $\tilde{K}$ over $\mathbb{Q}_p$ can be obtained as follows ($\tilde{\mathfrak{p}} = Jac(\tilde{R})$, $\tilde{R}$ maximal R-order in $\tilde{K}$): Choose α to be minimal with the above property and

let $H = \langle\sigma\rangle$ *be the subgroup of the Galois group of* $\mathbb{Q}_p[\zeta_\alpha]$ *over* $\mathbb{Q}_p$ *the fixed field of which is* $\tilde{K}$.

(i) *If* $|H| = 1$, *then* $\tilde{K} = \mathbb{Q}_p[\zeta_\alpha]$ *and* $d = d_\alpha := p^{\alpha-1}(\alpha p-\alpha-1)$

(ii) *If* p *is odd, then* $|H| = h \mid p-1$ *and*
$d = \frac{1}{h}(d_\alpha+1)-1 = \frac{1}{h}(p^{\alpha-1}(\alpha p-\alpha-1)+1)-1$

(iii) *If* $p = 2$ *and* $|H| \neq 1$, *then* $|H| = 2$ *and*
$d = \frac{1}{2} d_\alpha - 1 = 2^{\alpha-2}(\alpha-1)-1$

Proof: Clearly by elementary Galois theory all possible fields are taken care of in (i), (ii), and (iii). Note, the group $\langle\zeta_\alpha\rangle$ extended by H (with the natural action) is among the groups investigated in (III.14). In case of (i) and (ii), the (1,1)-entry of the exponent matrix of the kernel of $\psi_{\alpha,d}$ is equal to $\frac{p^{\alpha-1}-1}{h}+1$; namely the Cartan matrix of $FG_{\alpha-1,h}$ is easily seen to be $I_h + \frac{p^{\alpha-1}-1}{h} J_h$, where $I_h, J_h \in \mathbb{Z}^{h\times h}$ are the unit matrix and the matrix with all entries equal to one. Hence by the conductor formula (III.8) (actually Jacobinski's version suffices for this application) yields

$$\frac{p^{\alpha-1}-1}{h} + 1 = (\mu_s - d_s =) \frac{p - p^{\alpha-1}}{h} - d$$

and (i) and (ii) follow. (Note, (ii) also follows from (i) by observing that $\mathbb{Q}_p[\zeta_\alpha]$ is tamely ramified over $\tilde{K}$ and applying Dededinds different formula, cf. proof also of (iii). Note also that one can avoid to apply the conductor formula, if one uses the well known fact that the discriminant of a group ring RG with respect to the regular trace is equal to $|G|^{|G|}R$.) To prove (iii), note that $\mathbb{Q}_p[\zeta_\alpha]$ is a quadratic extension of $\tilde{K} = \mathbb{Q}_p[\zeta_\alpha+\sigma(\zeta_\alpha)]$, where $H = \langle\sigma\rangle$. The minimum polynomial of ζ_α over $\tilde{K}$ is $(x-\zeta_\alpha)(x-\sigma(\zeta_\alpha)) = x^2-sx+n$ with $s = \zeta_\alpha+\sigma(\zeta_\alpha)$ and $n = \zeta_\alpha\cdot\sigma(\zeta_\alpha)$. Hence the discriminant of $\mathbb{Q}_p[\zeta_\alpha]$ over $\tilde{K}$ is equal to $\det\begin{pmatrix} 2 & s \\ s & -2n+s^2\end{pmatrix} = s^2-4n$. The element s generates $\tilde{\mathfrak{p}}$ (note, $\sigma(\zeta_\alpha) = \zeta_\alpha^{1+2^{\alpha-1}}$ cannot occur because of the minimality of α).

Hence the discriminant generates $\tilde{\mathfrak{p}}^2$. Since $\mathbb{Q}_p[\zeta_\alpha]$ and $\tilde{K}$ are totaly ramified over $\mathbb{Q}_p$ one gets $d_\alpha = d\cdot 2+2$.

q.e.d.

It should be noted that the differents in case (iii) could be obtained by the conductor formula. However, in this case the index of $X_{\sigma,\alpha}$ in its graduated hull in (III.14) (iii) had to be computed, as indicated at the end of the proof there. In the two cases relevant for (III.15) (iii) $X_{\sigma,\alpha}$ is the unique maximal suborder of its graduated hull.

It is well known since Brauer that the degrees of the projective RG-lattices are divisible by the order of the p-Sylow subgroups of G where p is the characteristic of F . This result as well as analogous (known) congruences for the values of the characters of the projective lattices on the other conjugacy classes of G can be obtained from the following analysis of the center of a selfdual order.
Let $\omega_s : Z(A) \longrightarrow Z(A_s)$ denote the projection of the center of A onto the center of A_s which is identified with Z_s $(1 \leq s \leq h)$.
Recall the definition of the d'_{si} given before (III.11) (as the multiplicity of V_s in KP_i , $1 \leq i \leq r$, $1 \leq s \leq h$) .

(III.16) Proposition. For $1 \leq i \leq r$ and any z in the center of the selfdual order Λ one has

$$\sum_{s=1}^{h} m_s d'_{si}\, tr_{Z_s/K}(\omega_s(u\cdot z)) \in R$$

Proof: Let $\overset{\circ}{\varepsilon}_i$ be a primitive idempotent of Λ with $P_i \cong \Lambda\overset{\circ}{\varepsilon}_i$.
Then $T_u(\overset{\circ}{\varepsilon}_i z) \in R$ since $\overset{\circ}{\varepsilon}_i z \in \Lambda$. One has

$$T_u(\overset{\circ}{\varepsilon}_i z) = \sum_{s=1}^{h} tr_s(\overset{\circ}{\varepsilon}_i u_s z) = \sum_{s=1}^{h} tr_{Z_s/K}(tr_{A_s/Z_s}(\overset{\circ}{\varepsilon}_i \varepsilon_s u z)) =$$

$$= \sum_{s=1}^{h} tr_{Z_s/K}(m_s d'_{si}\, u_s z) = \sum_{s=1}^{h} m_s d'_{si}\, tr_{Z_s/K}(\omega_s(u\, z)) .$$

q.e.d.

In case $\Lambda = RG$ is the group ring of a finite group G (III.16) becomes $\chi_{P_i}(g) = \sum_{s=1}^{h} d'_{si} tr_{Z_s/K}(\chi_s(g))) \in |C_G(g)|R$ for all $g \in G$, where χ_{P_i} resp. χ_s is the K-character of G afforded by KP_i resp. V_s for $1 \leq i \leq r$, $1 \leq s \leq h$. This result is known as a consequence of the Green correspondence (in the splitting situation, which, however, implies the above result), cf. [Fei 82], page 144. There is

another aspect of (III.16), which might play a rôle when one analyses Λ under the assumption that the center $Z(\Lambda)$ (say character table in case of a group ring) and the decomposition numbers are already known. Namely the $\overset{o}{\varepsilon}_i \Lambda \overset{o}{\varepsilon}_i$ are again selfdual orders and one $\overset{o}{\varepsilon}_i Z(\Lambda) \overset{o}{\varepsilon}_i$ is isomorphic to the epimorphic image $\varepsilon_{c_i} Z(\Lambda)$ of $Z(\Lambda)$, where $\varepsilon_{c_i} = \sum_{d_{si} \neq 0} \varepsilon_s$. These two pieces of information might be a substantial help for the analysis.

IV. Selfdual orders with decomposition numbers 0 and 1

In this chapter R is a complete discrete valuation ring, K the field of fractions of R, $\mathfrak{p}$ the maximal ideal of R, and $F = R/\mathfrak{p}$ the residue class field. $A = \bigoplus_{s=1}^{h} A_s$ is a separable K-algebra with minimal twosided ideals $A_s = \varepsilon_s A \cong D_s^{l_s \times l_s}$ and central primitive idempotents ε_s for $s=1,\dots,h$, where $l_s \in \mathbb{N}$ and D_s is a K-division algebra with center Z_s and dimension m_s^2, $m_s \in \mathbb{N}$, over Z_s. The maximal R-order in D_s is denoted by Ω_s, the radical $\mathrm{Jac}(\Omega_s)$ by $\mathfrak{P}_s$, the residue class field $F_s = \Omega_s/\mathfrak{P}_s$ the residue degree $\dim_F F_s$ by f_s and the ramification index by e_s (i.e. $\mathfrak{p}\Omega_s = \mathfrak{P}_s^{e_s}$) for $1 \le s \le h$. Moreover Λ denotes an R-order in A, which is selfdual with respect to the generalized trace map $T_u: A \to K$ with $u = \sum_{s=1}^{h} u_s \in Z(A)$, $(u_s \in Z_s \setminus \{0\}, s=1,\dots,h)$, cf. Chapter III.
The components of $\Lambda/\mathrm{Jac}(\Lambda)$ are denoted by $\bar{\Lambda}_i$ with $1 \le i \le r$. Moreover $S_1,\dots,S_r$ denote (representatives of the isomorphism classes of the) simple Λ-(torsion)modules with $\bar{\Lambda}_i S_i = S_i$ $(1 \le i \le r)$. A projective cover of S_i is denoted by P_i, $1 \le i \le r$. Finally let $\tilde{\varepsilon}_1,\dots,\tilde{\varepsilon}_r$ be orthogonal idempotents of Λ such that $\tilde{\varepsilon}_i + \mathrm{Jac}(\Lambda) \in \bar{\Lambda}_i$ are the central primitive idempotents of $\Lambda/\mathrm{Jac}(\Lambda)$.
The basic assumption of this chapter is the following:
$\varepsilon_s\Lambda$ is a graduated R-order for each s, $1 \le s \le h$.

Note, Proposition (III.12) gives conditions for the $\varepsilon_s\Lambda$ to be graduated orders. In general this condition can be enforced by an unramified ground ring extension, if all (modified) decomposition numbers (cf. Definition (III.10)) are equal to 0 and 1. Afterwards the Galois descent can be performed by means of Theorem (II.20).

The assumption of this chapter that $\Gamma = \bigoplus_{s=1}^{h} \varepsilon_s \Lambda$ is a graduated R-order implies in particular that Γ is the unique graduated hull of Λ ; moreover, the conductor of Γ in Λ is given by Γ^* as described in (III.8). Somewhat more notation is needed. Let d_{si} denote the decomposition numbers of Λ for $1 \leq s \leq h$, $1 \leq i \leq r$, cf. (III.10),

$$r_s = \{i \mid 1 \leq i \leq r,\ d_{si} = 1\} \qquad \text{for} \quad 1 \leq s \leq h \text{ , and}$$

$$c_i = \{s \mid 1 \leq s \leq h,\ d_{si} = 1\} \qquad \text{for} \quad 1 \leq i \leq r \text{ .}$$

Then there exist exponent matrices $M^{(s)} = (m_{ij}^{(s)})_{i,j \in r_s} \in \mathbb{Z}_{\geq 0}^{|r_s| \times |r_s|}$

such that

$$\varepsilon_s \Lambda \cong \Lambda(\Omega_s, \tilde{n}^{(s)}, M^{(s)})$$

where $M^{(s)}$ satisfies (*) of Chapter IIa and is normalized in such a way that the first column is equal to zero. $\tilde{n}^{(s)}$ is given by $\tilde{n}^{(s)} = (\dim_{\Omega_s / \mathfrak{P}_s} S_i)_{i \in r_s} \in \mathbb{N}^{1 \times |r_s|} (1 \leq s \leq h)$. According to (****) of Chapter IIa the structural invariants of $\varepsilon_s \Lambda$ satisfy $m_{ijk}^{(s)} = m_{ij}^{(s)} + m_{jk}^{(s)} - m_{ik}^{(s)}$ and can be defined by

$$(\varepsilon_{si} \Lambda \varepsilon_{sj})(\varepsilon_{sj} \Lambda \varepsilon_{sk}) = \mathfrak{P}_s^{m_{ijk}^{(s)}} \varepsilon_{si} \Lambda \varepsilon_{sk} \quad \text{for} \quad i,j,k \in r_s \text{ ,}$$

where $\varepsilon_{si} = \varepsilon_s \cdot \tilde{\varepsilon}_i$.

An important consequence of $d_{si} \leq 1$ for $1 \leq s \leq h$ and $1 \leq i \leq r$ is obtained by applying Brauer reciprocity (III.11). For each projective indecomposable Λ-lattice P_i the A-module $KP_i := K \otimes_R P_i$ is isomorphic to the direct sum $\bigoplus_{s \in c_i} V_s$ of nonisomorphic irreducible A-modules:

$$KP_i = \bigoplus_{s \in c_i} \varepsilon_s KP_i \quad \text{with} \quad \varepsilon_s KP_i \cong V_s \quad \text{for} \quad s \in c_i \quad \text{and} \quad i=1,\dots,r \text{ .}$$

This will play a major rôle in this chapter later on.

The aim of this chapter is to derive equations for the structural invariants $m_{ijk}^{(s)}$ resp. for the exponent matrices $M^{(s)}$. In favourable situations these considerations suffice to determine the graduated hull Γ of Λ completely up to isomorphism. The first type of equations is very easily derived and consists of the equalities induced by automorphisms and antiautomorphisms of Λ. In case Λ is the group ring of a finite group G, i.e. $\Lambda = RG$ with R as in (III.2), one has the following obvious automorphisms:

(i) each automorphism α of G induces the automorphism

$\sum_{g \in G} a_g g \rightarrow \sum_{g \in G} a_g \alpha(g)$ of RG ($a_g \in R$) ;

(ii) each homomorphism χ of G into the unit group of R (linear character) induces the automorphism

$\sum_{g \in G} a_g g \rightarrow \sum_{g \in G} \chi(g) a_g g$ of RG .

Moreover, the following antiautomorphism of RG is well known:

(iii) $\sum_{g \in G} a_g g \rightarrow \sum_{g \in G} a_g g^{-1}$.

Under obvious conditions one can construct automorphisms and antiautomorphisms of the selfdual R-order εRH constructed in (III.4) in a similar way.

Clearly, each automorphism resp. antiautomorphism α of Λ induces an automorphism resp. antiautomorphism of $\Lambda/\mathrm{Jac}(\Lambda)$ and of A, and hence, α induces a permutation $\underline{\alpha}$ of the indices $1,\dots,r$ of the components of $\Lambda/\mathrm{Jac}(\Lambda)$ and a permutation $\bar{\alpha}$ of the indices $1,\dots,h$ of the components of A.

__(IV.1) Proposition.__ *(i) Let α be an automorphism of Λ. Then the structural invariants of $\varepsilon_s\Lambda$ and $\varepsilon_{\bar{\alpha}(s)}\Lambda$ are interrelated by*

$$m_{ijk}^{(s)} = m_{\underline{\alpha}(i)\underline{\alpha}(j)\underline{\alpha}(k)}^{(\bar{\alpha}(s))} \quad \text{for } i,j,k \in r_s,\ 1 \leq s \leq h .$$

(ii) Let α *be an antiautomorphism of* Λ . *Then the structural invariants of* $\varepsilon_s\Lambda$ *and* $\varepsilon_{\bar{\alpha}(s)}\Lambda$ *are interrelated by*

$$m_{ijk}^{(s)} = m_{\underline{\alpha}(k)\underline{\alpha}(j)\underline{\alpha}(i)}^{(\bar{\alpha}(s))} \quad \text{for} \quad i,j,k \in r_s,\ 1 \leq s \leq h .$$

Proof. After modifying α by a suitable inner automorphism of Λ one can assume $\alpha(\tilde{\varepsilon}_i) = \tilde{\varepsilon}_{\alpha(i)}$ for $i = 1,\ldots,r$. If α denotes also its extension to A , then $\alpha(\varepsilon_{si}) = \varepsilon_{\bar{\alpha}(s)\underline{\alpha}(i)}$ for $i \in r_s$, $s=1,\ldots,h$. The results follow now by applying α to the defining equation for $m_{ijk}^{(s)}$ above.

q.e.d.

Note, by virtue of equation (****) of Chapter IIa, one can easily rewrite the equations in (IV.1) into equations for the coefficients $m_{ij}^{(s)}$ of the exponent matrices $M^{(s)}$ of $\varepsilon_s\Lambda$. It will turn out in the case of $\Lambda = RG$ and the involution α described in (iii) above that the equations in (IV.1)(ii) are particularly helpful if $\bar{\alpha}(s) = s$ and $\underline{\alpha}(i) = i$ for all $i \in r_s$. (This means in terms of characters that the involved Frobenius and Brauer characters are real, i.e. have the same value on g and g^{-1} for all $g \in G$.) Namely in this case $M^{(s)}$ is completely determined by the $\frac{1}{2}|r_s|(|r_s|-1)$ entries above the diagonal:

$m_{ij}^{(s)} = m_{ji}^{(s)} + m_{kj}^{(s)} - m_{ki}^{(s)}$ where $i,j,k \in r_s$ and k is the index of the first column of $M^{(s)}$ (which is zero). This follows from the equations in (IV.1)(ii) and $m_{ij}^{(s)} = m_{ijk}^{(s)}$ (cf. (II.6)(ii) and the subsequent remarks in Chapter II).

There are two possible ways to express the following basic ideas of this paper for finding the graduated hull of Λ :
One can either work with projective indecomposable Λ-lattices and view them as subdirect products (or amalgams) of irreducible Λ-lattices, or

one can express Λ as a subdirect product of the Λ-Λ-bilattices $\varepsilon_s\Lambda$, $s = 1,\dots,h$. It seems to be worthwhile to present both formulations and give a dictionary for translating one into the other. For the projective indecomposable Λ-lattices $P_1,\dots,P_r$ one has

$$(\#) \quad \bigoplus_{s\in c_i} (\varepsilon_s P_i \cap P_i) \subseteq P_i \subseteq \bigoplus_{s\in c_i} \varepsilon_s P_i$$

The $\varepsilon_s P_i$ as well as the $\varepsilon_s P_i \cap P_i$ for $s \in c_i$ form irreducible Λ-lattices. $\bigoplus_{s\in c_i} \varepsilon_s P_i$ is the unique minimal Λ-lattice in KP_i $(:= K\otimes_R P_i)$ which is completely decomposable and contains P_i. Dually, $\bigoplus_{s\in c_i} (\varepsilon_s P_i \cap P_i)$ is the unique maximal Λ-sublattice of P_i which is completely decomposable. On the other side one has

$$(\#\#) \quad \bigoplus_{s=1}^{h} (\varepsilon_s\Lambda \cap \Lambda) \subseteq \Lambda \subseteq \bigoplus_{s=1}^{h} \varepsilon_s\Lambda$$

and similar statements can be made about the involved Λ-Λ-lattices. But $\Gamma = \bigoplus_{s=1}^{h} \varepsilon_s\Lambda$ is not only a Λ-Λ-lattice, but also the unique graduated hull of Λ and $\bigoplus_{s=1}^{h} (\varepsilon_s\Lambda \cap \Lambda)$ is easily seen to be the conductor of Γ in Λ, and hence, equal to Γ^* by (III.7). Viewing the modules in (##) only as (left) Λ-lattices and multiplying with a primitive idempotent from the right one is lead to (#). Applying the conductor formula (III.8) yields the following proposition. (For the irreducible lattices of graduated orders in standard form the terminology introduced in (II.4) is used.)

(IV.2) Proposition. *For $i=1,\dots,r$ and $s \in c_i$ the $\varepsilon_s\Lambda$-lattices $\varepsilon_s P_i|_{\varepsilon_s\Lambda}$ resp. $(\varepsilon_s P_i \cap P_i)|_{\varepsilon_s\Lambda}$ are projective resp. injective indecomposable $\varepsilon_s\Lambda$-lattices. If $\varepsilon_s\Lambda$ is identified with $\Lambda(\Omega_s, \tilde{n}^{(s)}, M^{(s)})$, they correspond to $L(M_i^{(s)})$ resp. $L({}_iM^{(s)})\mathfrak{P}_s^{\varkappa_s} = L({}_iM^{(s)} + \underbrace{(\varkappa_s,\dots,\varkappa_s)}_{|r_s|}{}^{tr})$, where $\varkappa_s$ is defined as in (III.8) and $M_i^{(s)}$, ${}_iM^{(s)}$ in (II.4).*

Note, (IV.2) allows to read off the Λ-composition factors of $\varepsilon_s P_i/(\varepsilon_s P_i \cap P_i)$ like (III.8) allows to read of the Λ-Λ-composition factors of Γ/Γ^*. The Λ-Λ-modules Γ/Λ and Λ/Γ^*, which measure how close ${}_\Lambda\Lambda_\Lambda$ comes to be the direct sum of irreducible Λ-Λ-lattices, give some insight into what the graduated hull Γ might look like. On the side of the Λ-lattices one is lead to the next definition.

(IV.3) Definition. *(i) For a Λ-lattice L the Λ-(torsion)modules $a_u(L) = (\bigoplus_{s=1}^{h} \varepsilon_s L)/L$ and $a_l(L) = L/\bigoplus_{s=1}^{h} (\varepsilon_s L \cap L)$ are called upper and lower amalgamating factor (module) of L.*

(ii) For a (nonempty) subset x of $\{1,\dots,h\}$ the central idempotent ε_x of A is defined by $\varepsilon_x = \sum_{s\in x} \varepsilon_s$.

(iii) For a partition σ of $\{1,\dots,h\}$ and any Λ-lattice L the Λ-(torsion)module $a_u^\sigma(L) = (\bigoplus_{x\in\sigma} \varepsilon_x L)/L$ resp. $a_l^\sigma(L) = L/\bigoplus_{x\in\sigma} (\varepsilon_x L \cap L)$ is called upper resp. lower amalgamating σ-factor of L.

Note, the isomorphism type of $a_u(L)$ and $a_l(L)$ as well as $a_u^\sigma(L)$ and $a_l^\sigma(L)$ only depend on σ and the isomorphism type of L, cf. [Ple 78], [Ple 80a], [Ple 80b] as detailed references. Note also, if $\sigma = \{\{1\},\dots,\{h\}\}$, then $a_u^\sigma(L) = a_u(L)$ and $a_l^\sigma(L) = a_l(L)$. The next theorem discusses the upper and lower amalgamating factors of the projective indecomposable Λ-lattices $P_1,\dots,P_r$ parallel to the Λ-Λ-modules Γ/Λ and Λ/Γ^* in (IV.5).

(IV.4) Theorem. *(i) The upper amalgamating factor $a_u(P_i) = (\bigoplus_{s\in c_i} \varepsilon_s P_i)/P_i$ has simple socle $a_u'(P_i) \cong S_i$ $(1 \leq i \leq r)$. The same holds for $a_u^\sigma(P_i)$ for any partition σ of $\{1,\dots,h\}$ with $c_i \nsubseteq x$ for all $x \in \sigma$.*

(i*) *The lower amalgamating factor* $a_l(P_i) = P_i / \bigoplus_{s\in c_i} (\varepsilon_s P_i \cap P_i)$ *has exactly one maximal* Λ*-submodule* $a_l'(P_i)$. *One has* $a_l(P_i)/a_l'(P_i) \cong S_i$ $(1 \leq i \leq r)$. *The same holds for* $a_l^\sigma(P_i)$ *for any partition* σ *of* $\{1,\dots,h\}$ *with* $c_i \nsubseteq x$ *for all* $x \in \sigma$.

(ii) *For each* $s \in c_i$ *there is a* Λ*-monomorphism of* $\varepsilon_s P_i/\varepsilon_s P_i \cap P_i$ *into* $a_u(P_i)$, $(1 \leq i \leq r)$. *More generally, for each partition* σ *of* $\{1,\dots,h\}$ *there is a monomorphism of* $\varepsilon_x P_i/\varepsilon_x P_i \cap P_i$ *into* $a_u^\sigma(P_i)$ *for every* $x \in \sigma$, $(x \cap c_i \neq \emptyset,\ x \nsupseteq c_i)$.

(ii*) *For each* $s \in c_i$ *there is a* Λ*-epimorphism of* $a_l(P_i)$ *onto* $\varepsilon_s P_i/\varepsilon_s P_i \cap P_i$, $(1 \leq i \leq r)$. *More generally, for each partition* σ *of* $\{1,\dots,h\}$ *there is a* Λ*-epimorphism of* $a_l(P_i)$ *onto* $\varepsilon_x P_i/\varepsilon_x P_i \cap P_i$ *for every* $x \in \sigma$, $(x \cap c_i \neq \emptyset$, $x \nsupseteq c_i)$.

(iii) *For* $s \in c_i$ *and* $1 \leq i \leq r$ *the following four* Λ*-(torsion) modules are isomorphic:* $\varepsilon_s P_i/\varepsilon_s P_i \cap P_i \cong (1-\varepsilon_s)P_i/(1-\varepsilon_s)P_i \cap P_i \cong$ $\cong \varepsilon_s P_i \oplus (1-\varepsilon_s)P_i/P_i \cong P_i/(P_i \cap \varepsilon_s P_i) \oplus (P_i \cap (1-\varepsilon_s)P_i)$. *The same statement holds if* s *is replaced by a nontrivial subset* x *of* c_i .

(iv) *(Reciprocity) The multiplicity of* S_j *in (a* Λ*-composition series of)* $a_u(P_i)$ *is equal to the multiplicity of* S_i *in* $a_l(P_j)$, $1 \leq i, j \leq r$. *More generally, let* σ *and* τ *be partitions of* $\{1,\dots,h\}$ *with* σ *finer than* τ . *Then there are* Λ*-monomorphisms* $\alpha_i\colon a_u^\tau(P_i) \to a_u^\sigma(P_i)$ *and* Λ*-epimorphisms* $\beta_j\colon a_l^\sigma(P_j) \to a_l^\tau(P_j)$ *such that the multiplicity of* S_j *in* $\mathrm{coker}(\alpha_i)$ *is equal to the multiplicity of* S_i *in* $\ker(\beta_j)$ *for* $1 \leq i, j \leq r$.

(v) *Let* μ_{ij} *resp.* λ_{ij} *be the multiplicity of* S_i *in* $a_u(P_j)$ *resp.* $a_l(P_j)$ *for* $1 \leq i, j \leq r$, *then*

$$\mu_{ij} + \lambda_{ij} = \sum_{s\in c_j} (\varkappa_s - m_{ij}^{(s)} - m_{ji}^{(s)}) = \sum_{s\in c_j} (\varkappa_s - m_{iji}^{(s)})$$ *with* $\varkappa_s$ *as defined in (III.8).*

(vi) $\mu_{ij} \leq (|c_j|-1)\lambda_{ij}$ *and* $\lambda_{ij} \leq (|c_j|-1)\mu_{ij}$ *for* $1 \leq i, j \leq r$.

Some notation is needed for the bimodule version of this theorem. Like in (III.9) M_{ij} denotes a simple Λ-Λ-module with $\bar{\Lambda}_i M_{ij} \bar{\Lambda}_j = M_{ij}$ for $1 \leq i, j \leq r$. A partition σ of $\{1,\ldots,h\}$ defines an Λ-Λ-bimodule (or even an R-order) $\Lambda_\sigma = \bigoplus_{x \in \sigma} \varepsilon_x \Lambda$. Clearly, $\Lambda_\sigma^* = \bigoplus_{x \in \sigma} (\varepsilon_x \Lambda \cap \Lambda)$, and $\Lambda_\tau = \Gamma$ for $\tau = \{\{0\},\ldots,\{h\}\}$. For brevity's sake the most important special cases are not reformulated as in (IV.4).

<u>*(IV.5) Theorem.*</u> *Let σ be a partition and x a subset of $\{1,\ldots,h\}$ and let $s(\sigma) = \{i \mid 1 \leq i \leq r,\ c_i \not\subseteq y$ for all $y \in \sigma\}$.*

(i) The socle of the Λ-Λ-module Λ_σ/Λ is isomorphic to $\bigoplus_{i \in s(\sigma)} M_{ii}$.

(i^*) The head of the Λ/Λ_σ^* is isomorphic to $\bigoplus_{i \in s(\sigma)} M_{ii}$.

(ii) If $x \in \sigma$, there is a Λ-Λ-monomorphism of $\varepsilon_x\Lambda/\varepsilon_x\Lambda \cap \Lambda$ into Λ_σ/Λ.

(ii^*) If $x \in \sigma$, there is a Λ-Λ-epimorphism of Λ/Λ_σ^* onto $\varepsilon_x\Lambda/\varepsilon_x\Lambda \cap \Lambda$.

(iii) The following four Λ-Λ-(torsion)modules are isomorphic:

$\varepsilon_x\Lambda/\varepsilon_x\Lambda \cap \Lambda \cong (1-\varepsilon_x)\Lambda \,/\, (1-\varepsilon_x)\Lambda \cap \Lambda \cong \varepsilon_x\Lambda \oplus (1-\varepsilon_x)\Lambda/\Lambda \cong \Lambda/\varepsilon_x\Lambda \cap \Lambda \oplus (1-\varepsilon_x)\Lambda \cap \Lambda$

(iv) (Reciprocity) Let τ be a second partition of $\{1,\ldots,h\}$ such that σ is finer than τ. Then $\Lambda_\sigma \supseteq \Lambda_\tau$ and the multiplicity of M_{ji} in $\Lambda_\sigma/\Lambda_\tau$ is equal to the multiplicity of M_{ij} in $\Lambda_\tau^*/\Lambda_\sigma^*$, and both are equal to the multiplicities occurring in the last part of (IV.4)(iv), for $1 \leq i, j \leq r$.

(v) Let μ_{ij} and λ_{ij} be the multiplicities defined in (IV.4)(v) for $1 \leq i, j \leq r$. Then μ_{ij} and λ_{ij} are equal to the multiplicities of M_{ij} in Γ/Λ and Λ/Γ^* respectively.

Proof of (IV.4) and (IV.5): (i) and (i^*). P_i and all its factor modules $\neq 0$ have exactly one maximal submodule. The corresponding factor module is isomorphic to S_i. But $a_1^{\sigma}(P_i)$ is a factor module of P_i; moreover, $a_1^{\sigma}(P_i) \neq 0$ iff $c_i \nsubseteq x$ for any $x \in \sigma$. This proves (IV.4)(i*). Statement (IV.5)(i*) follows from this and the corresponding version for right projective indecomposables. Namely ${}_{\Lambda}\Lambda$ splits into a direct sum of indecomposables, a_1^{σ} commutes with direct sums. Let k_{σ} be the preimage of the intersection of the maximal submodules of $a_1^{\sigma}({}_{\Lambda}\Lambda) = {}_{\Lambda}(\Lambda/\Lambda_{\sigma}{}^*)$ under the natural epimorphism of ${}_{\Lambda}\Lambda$ onto $a_1^{\sigma}({}_{\Lambda}\Lambda)$. The same construction for Λ_{Λ} leads to the same k_{σ}; hence k_{σ} is Λ-Λ-submodule of ${}_{\Lambda}\Lambda_{\Lambda}$ and the head of Λ/Λ_{σ} is isomorphic to Λ/k_{σ}. Since Λ/k_{σ} must be isomorphic to a Λ-Λ-submodule of $\Lambda/\mathrm{Jac}(\Lambda) \cong \bigoplus_{i=1}^{r} M_{ii}$, (IV.4)(i*) now implies (IV.5)(i*). Furthermore (IV.5)(i) follows from (IV.5)(i*); namely the socle of Λ_{σ}/Λ is isomorphic to $k_{\sigma}{}^*/\Lambda$. (cf. also the proof of (III.9)). Finally (IV.4)(i) follows from (IV.5)(i). To prove this, let $\mathring{\varepsilon}_i$ be a primitive idempotent of Λ with $P_i \cong \Lambda\mathring{\varepsilon}_i$. Then $a_u^{\sigma}(P_i) \cong \Lambda_{\sigma}\mathring{\varepsilon}_i/\Lambda\mathring{\varepsilon}_i \cong (\Lambda_{\sigma}/\Lambda)\mathring{\varepsilon}_i$ and the socle of $a_u^{\sigma}(P_i)$ is isomorphic to $(k_{\sigma}^*/\Lambda)\mathring{\varepsilon}_i$ $(1 \leq i \leq r)$. (IV.4)(i) follows.

(ii), (ii*), and (iii). One has the following chain of Λ-lattices:

$$\bigoplus_{y\in\sigma} \varepsilon_y P_i \cap P_i \subseteq (\varepsilon_x P_i \cap P_i) \oplus ((1-\varepsilon_x)P_i \cap P_i) \subseteq P_i \subseteq \varepsilon_x P_i \oplus (1-\varepsilon_x)P_i \subseteq$$
$$\subseteq \bigoplus_{y\in\sigma} \varepsilon_y P_i$$

for $1 \leq i \leq r$, since $x \in \sigma$. Hence $(\varepsilon_x P_i \oplus (1-\varepsilon_x)P_i)/P_i$ embeds into $a_u^{\sigma}(P_i)$ and $P_i/(\varepsilon_x P_i \cap P_i) \oplus ((1-\varepsilon_x)P_i \cap P_i)$ is an epimorphic image of $a_1^{\sigma}(P_i)$. Clearly the composition of the epimorphism $P_i \to \varepsilon_x P_i : \alpha \to \varepsilon_x\alpha$ and the natural epimorphism $\varepsilon_x P_i \to \varepsilon_x P_i/(\varepsilon_x P_i \cap P_i)$ has kernel $((1-\varepsilon_x)P_i \cap P_i) \oplus (\varepsilon_x P_i \cap P_i)$. The same holds if ε_x is replaced by $1 - \varepsilon_x$. This, together with the isomorphism theorem (see fig. 1), implies (IV.4)(iii) (Compare also with diagram (***) on page 195 in [Ple 78]). By the earlier remarks (IV.4)(ii) and (ii*) follow.

The proof of the corresponding statements of (IV.5) is completely analogous.

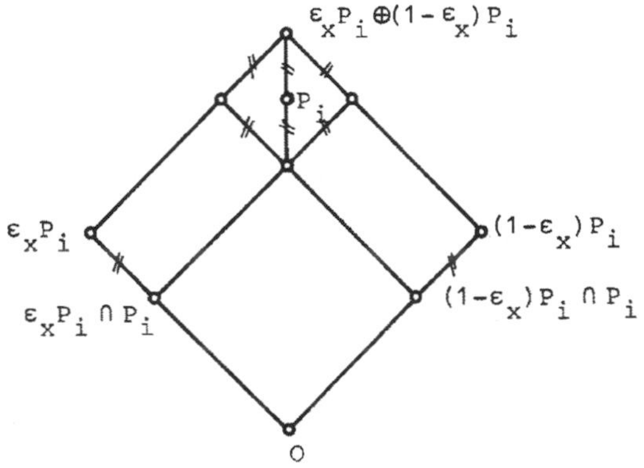

fig. 1

(iv) $\Lambda_\sigma \supseteq \Lambda_\tau$ is obvious; the other statements of (IV.5)(iv) are consequences of (III.9). The version of (IV.4) follows in the usual manner by multiplying with a primitive idempotent of Λ from the right side.

(v) Since $\lambda_{ij} + \mu_{ij}$ is equal to the multiplicity of S_i in $\bigoplus_{s \in c_i} (\varepsilon_s P_j / \varepsilon_s P_j \cap P_j)$, the statement of (IV.4) follows from (IV.2). The corresponding statement of (IV.5) follows from (III.8).

(vi) For $s \in c_j$ let v_s be the multiplicity of S_i in $\varepsilon_s P_j / (\varepsilon_s P_j \cap P_j)$ for fixed i, j with $1 \leq i, j \leq r$. By (IV.4)(v) one has $\sum_{s \in c_j} v_s = \mu_{ij} + \lambda_{ij}$. Since $\varepsilon_s P_j / (\varepsilon_s P_j \cap P_j)$ is submodule of $a_u(P_j)$ and factor module $a_l(P_j)$ by (IV.4)(ii) and (ii*), one has $v_s \leq \mu_{ij}$ and $v_s \leq \lambda_{ij}$ for all $s \in c_j$. Hence, $|c_j| \mu_{ij} \geq \mu_{ij} + \lambda_{ij}$ and $|c_j| \lambda_{ij} \geq \mu_{ij} + \lambda_{ij}$, which implies the claimed inequalities.

q.e.d.

For actual computations it is convenient to have a clear book-keeping device for the contributions of the $\varepsilon_s P_i/(\varepsilon_s P_i \cap P_i)$ $(1 \leq i \leq r, s \in c_i)$ to the amalgamating factors $a_l(P_i)$ and $a_u(P_i)$. This is provided by the "amalgamation matrix".

(IV.6) Definition. *For each projective indecomposable Λ-lattice P_i $(1 \leq i \leq r)$ the amalgamation matrix $\alpha(P_i) \in \mathbf{Z}_{\geq 0}^{|c_i| \times r}$ is defined as follows: The rows are indexed by the elements of c_i, the columns by $1,\dots,r$. The entry in the (s,j)-position $(1 \leq j \leq r,\ s \in c_i)$ is the multiplicity of S_j in $\varepsilon_s P_i/(\varepsilon_s P_i \cap P_i)$, namely $\varkappa_s - m_{ij}^{(s)} - m_{ji}^{(s)} = \varkappa_s - m_{iji}^{(s)}$ (cf. (IV.2)) in case $j \in r_s$ and 0 otherwise.*

In the next chapters many examples of amalgamation matrices can be found. Note, if α_{isj} denotes the entry in the (s,j)-position of $\alpha(P_i)$, then $\mathrm{Hom}_\Lambda(P_j, \varepsilon_s P_i/(\varepsilon_s P_i \cap P_i))$ is isomorphic to $\Omega_s/\mathfrak{P}_s^{\alpha_{isj}}$ as R-module. The simplest applications of (IV.4) are given in the following corollary.

(IV.7) Corollary. *Let $1 \leq i, j \leq r$.*

(i) If $i \neq j$ and $c_i \cap c_j$ consists of exactly one element, say s, then $m_{ij}^{(s)} + m_{ji}^{(s)} = \varkappa_s$.

(ii) If $i \neq j$ and $c_i \cap c_j$ consists of exactly two elements, say s and t, then $\varkappa_s - m_{ij}^{(s)} - m_{ji}^{(s)} = \varkappa_t - m_{ij}^{(t)} - m_{ji}^{(t)}$. Moreover, if the two ramification indices e_s and e_t of R in Ω_s resp. Ω_t are not equal, then $\varkappa_s - m_{ij}^{(s)} - m_{ji}^{(s)} \leq \min(e_s, e_t)$.

(iii) Let $\tau_{ij}^{(s)}$ be the smallest integer greater or equal than the quotient of the (s,j)-entry of the amalgamation matrix $\alpha(P_i)$ divided by the ramification index e_s of R in Ω_s. Then the two biggest

numbers $\tau_{ij}^{(s)}$ *for fixed* $1,j \in \{1,\dots,r\}$ *and all* $s \in c_i$ *are equal.*

Proof: (i) Note, $c_i \cap c_j = \{s\}$ means that the only entry in the j-th column of the amalgamation matrix $\alpha(P_i)$ which might possibly be unequal to zero is in the s-th row. By (IV.4) one has $\varepsilon_s P_i/\varepsilon_s P_i \cap P_i \cong (1-\varepsilon_s)P_i/(1-\varepsilon_s)P_i \cap P_i$. Since $c_i \cap c_j = \{s\}$, S_j cannot occur as composition factor in $(1-\varepsilon_s)P_i/(1-\varepsilon_s)P_i \cap P_i$ (note $(1-\varepsilon_s)P_i = (\sum_{\substack{t \neq s \\ t \in c_i}} \varepsilon_t)P_i)$; hence S_j can also not occur as composition factor in $\varepsilon_s P_i/\varepsilon_s P_i \cap P_i$, i.e. the entry in the (s,j)-position of $\alpha(P_i)$, namely $\varkappa_s - m_{ij}^{(s)} - m_{ji}^{(s)}$, is equal to zero.

(ii) Let $\varepsilon = 1 - \varepsilon_s - \varepsilon_t$. Then $\mathrm{Hom}_\Lambda(P_j, \varepsilon_s P_i \oplus \varepsilon P_i/(1-\varepsilon_t)P_i) = 0$, since $(1-\varepsilon_t)P_i$ is a subdirect product of $\varepsilon_s P_i$ and εP_i , and S_j does not occur as composition factor of Λ-factor modules of εP_i , which are R-torsion modules. (Note, in the case of sub-direct products of two lattices L_1 and L_2 the upper and lower amalgamating factors are both isomorphic to the same factor modules of L_1 and of L_2 , cf. (IV.4)(iii) ore more generally (***) on page 195 in [Ple 78].) The same holds, if ε_s and ε_t are interchanged. Hence, by (IV.4):

$$\mathrm{Hom}_\Lambda(P_j, \varepsilon_s P_i/\varepsilon_s P_i \cap P_i) \cong \mathrm{Hom}_\Lambda(P_j, \varepsilon_s P_i \oplus (1-\varepsilon_s)P_i/P_i) \cong$$
$$\cong \mathrm{Hom}_\Lambda(P_j, \varepsilon_s P_i \oplus \varepsilon_t P_i \oplus \varepsilon P_i/P_i) \cong \mathrm{Hom}_\Lambda(P_j, \varepsilon_t P_i \oplus (1-\varepsilon_t)P_i/P_i) \cong$$
$$\cong \mathrm{Hom}_\Lambda(P_j, \varepsilon_t P_i/\varepsilon_t P_i \cap P_i) \quad (\cong \mathrm{Hom}_\Lambda(P_j, X/P_i) \text{ with}$$
$$X = (\varepsilon_s P_i \oplus (1-\varepsilon_s)P_i) \cap (\varepsilon_t P_i \oplus (1-\varepsilon_t)P_i) \text{ cf. fig. 2)}.$$

By the remarks preceding this corollary the statement follows from this chain of R-isomorphisms.

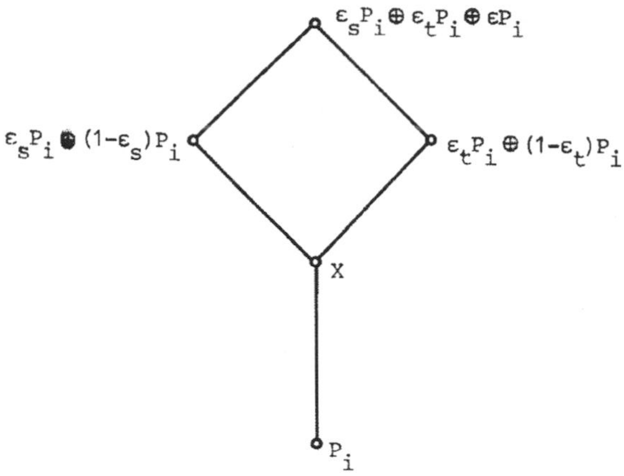

fig. 2

(iii) $\mathfrak{p}_{ij}^{\tau(s)}$ is the annihilator of $\mathrm{Hom}_\Lambda(P_j, \varepsilon_s P_i/\varepsilon_s P_i \cap P_i)$ in R. The statement now follows easily, since by (IV.4)(ii)(ii*), and (iii) $\varepsilon_s P_i/\varepsilon_s P_i \cap P_i$ is isomorphic to a section of $\bigoplus_{\substack{t \in C_i \\ t \neq s}} \varepsilon_t P_i/\varepsilon_t P_i \cap P_i$ for each $\varepsilon \in C_i$.

q.e.d.

The reciprocity statement (iv) in Theorem (IV.5) can be improved and better understood in the case of bimodules. If M is a Λ-Λ-module consisting of R-torsion elements, then $\mathrm{Hom}_R(M, K/R)$ can be turned into a Λ-Λ-module M^* by defining $(l_1 \varphi l_2)(m) = \varphi(l_2 m l_1)$ for all $\varphi \in M^*$, $l_1, l_2 \in \Lambda$, $m \in M$. Moreover, if N is another Λ-Λ-module which is an R-torsion module and $\Phi: M \to N$ a Λ-Λ-homomorphism, then $\Phi^*: N^* \to M^*$ defined by $\Phi^*(\varphi) = \varphi\Phi$ for all $\varphi \in N^*$ is a Λ-Λ-homomorphism. Since K/R is injective as R-module cf. [Rot 79], * a contravariant exact endofunctor for the Λ-Λ-modules which are R-torsion modules. Moreover, the composition of * with itself yields an endofunctor which is naturally equivalent to the identity functor. For the simple Λ-Λ-modules one has $M_{ij}^* \cong M_{ji}$ for $1 \leq i, j \leq r$. The refinement of (IV.5)(iv) then reads as follows.

<u>*(IV.8) Remark.*</u> *Let* σ *and* τ *be partitions of* $\{1,\ldots,h\}$ *with* σ *finer than* τ. *Then* $\Lambda_\sigma \supseteq \Lambda_\tau$ *and* $(\Lambda_\sigma/\Lambda_\tau)^* \cong \Lambda_\tau^*/\Lambda_\sigma^*$. *In particular* $(\Gamma/\Lambda)^* \cong \Lambda/\Gamma^*$.

Proof: Let $\tilde{T}_u$ denote the composition of T_u and the natural R-homomorphism $K \to K/R$. Then $\Lambda_\tau^* \to (\Lambda_\sigma/\Lambda_\tau) : l \to \varphi_l$ with $\varphi_l(x+\Lambda_\tau) = \tilde{T}_u(lx)$ for all $x \in \Lambda_\sigma$ defines a Λ-Λ-epimorphism with kernel Λ_σ^*.

q.e.d.

Unfortunately, the above procedure does not work for left Λ-modules, since they become right Λ-modules by the analogous procedure. This shortcoming can be overcome if Λ has an antiautomorphism. Therefore assume for the subsequent discussion that $\alpha:\Lambda \to \Lambda$ is a fixed anti-automorphism of Λ. A right Λ-module M can be turned into a left Λ-module M^α by defining $lm := m\alpha^{-1}(l)$ for $l \in \Lambda$, $m \in M$. One easily checks that this procedure defines a covariant exact functor α from the category of right Λ-modules to that of left Λ-modules; namely for a homomorphism $\Phi: M_1 \to M_2$ of right Λ-modules one defines $\Phi^\alpha : M_1^\alpha \to M_2^\alpha : m_1 \to \Phi(m_1)$. For A-modules one proceeds analogously. On the other hand the left Λ-modules M (resp. A-modules), which are of interest here, define right Λ-modules M^* as follows:
(α) If M is a left Λ-module which is an R-torsion module, then $M^* = \mathrm{Hom}_R(M,K/R)$; (β) if M is a left Λ-lattice, then $M^* = \mathrm{Hom}_R(M,R)$; (γ) if M is a left A-module, then $M^* = \mathrm{Hom}_K(M,K)$. In all three cases the right Λ- (resp. A-) module structure is imposed on M^* by $(\varphi l)(m) = \varphi(lm)$ for all $\varphi \in M^*$,

$m \in M$ and $l \in \Lambda$ (resp. $l \in A$) . There might be some confusion of the definition of M^* in case M is a left Λ-lattice contained in A with the earlier definition of M^* by means of T_u in Chapter III. Note, however, if $KM = A$ in this case, there is an obvious Λ-isomorphism between $\{x \in A \mid T_u(xM) \subseteq R\}$ and $\mathrm{Hom}_R(M,R)$, due to the fact that T_u induces a nondegenerate bilinear form on A . In case $KM \neq A$ always $M^* = \mathrm{Hom}_R(M,R)$. Returning to a general module M , the desired procedure, how to impose a new right module structure on M , consists in forming $M^\# := (M^*)^\alpha$. One easily checks $S_i^\# = S_{\underline{\alpha}(i)}$ $(1 \leq i \leq r)$ for the simple Λ-modules, and $V_s^\# = V_{\bar{\alpha}(s)}$ $(1 < s < h)$ for the irreducible A-modules (in the terminology introduced before (IV.1)). Again one easily sees that $\#$ defines a contravariant exact endofunctor of the category of right Λ-modules which are R-torsion modules, namely the composition of two exact functors. The important properties of $\#$ are the following two, only the second one of which uses the property of Λ to be selfdual.

(IV.9) Proposition. *(i) Let L be a Λ-lattice. Then the A-modules $(KL)^\#$ and $KL^\#$ are isomorphic (and will be identified). There is a duality $\delta : \mathfrak{Z}(LK) \to \mathfrak{Z}(KL^\#) : M \to \delta(M) := \{\varphi \in KL^\# \mid \varphi(M) \subseteq R\}$, where $\mathfrak{Z}(V) = \{L \subseteq V \mid L$ is Λ-lattice with $KL = V\}$ for any A-module V . One has*

(α) $\delta(M) \cong M^\#$ for all $M \in \mathfrak{Z}(KL)$;

(β) $(M_1/M_2)^\# \cong \delta(M_2)/\delta(M_1)$ for $M_1, M_2 \in \mathfrak{Z}(KL)$, $M_1 \subseteq M_2$;

(ii) If P is a projective Λ-lattice, the same holds for $P^\#$. Moreover, $P_i^\# \cong P_{\underline{\alpha}(i)}$ for $1 \leq i \leq r$.

Proof. (i) Straightforward by choosing compatible bases for L and M , resp. M_1 and M_2 and taking dual bases.

(ii) The nondegeneracy of the bilinear form induced by T_u on Λ implies that P_i^* is a right projective Λ-lattice. Namely, let $\overset{\circ}{\varepsilon}_i$ be a primitive idempotent of Λ with $\Lambda\overset{\circ}{\varepsilon}_i \cong P_i$. Then $\Phi : \overset{\circ}{\varepsilon}_i\Lambda \to \mathrm{Hom}_R(\Lambda\overset{\circ}{\varepsilon}_i, R)$ with $\Phi(x)(y) = T_u(xy)$ for $x \in \overset{\circ}{\varepsilon}_i\Lambda$ and $y \in \Lambda\overset{\circ}{\varepsilon}_i$ is an isomorphism of right Λ-lattices. (In particular that Λ is selfdual implies that Λ is a Gorenstein order, cf. [CuR 81], which, however, could already be seen from $\Lambda^* = \Lambda$ and the remarks preceding this proposition.) Finally $(\overset{\circ}{\varepsilon}_i\Lambda)^{\alpha} \to \Lambda\alpha(\overset{\circ}{\varepsilon}_i) : x \to \alpha(x)$ is an isomorphism of left Λ-lattices. Hence $P_i^* \cong P_{\underline{\alpha}(i)}$.

q.e.d.

The next theorem interrelates certain sections of upper and lower amalgamating factors of the projective indecomposables and compares multiplicities of the simple modules in these sections.

<u>*(IV.10) Theorem.*</u> *Let $\alpha : \Lambda \to \Lambda$ be an antiautomorphism of Λ and let σ, τ be partitions of $\{1,\ldots,r\}$ with σ finer than τ.*

(i) The following Λ-modules are isomorphic:

$$a_u(P_i)^{\#} \cong a_l(P_i^{\#}) \cong a_l(P_{\underline{\alpha}(i)}) \quad \text{for} \quad 1 \le i \le r .$$

More precisely, one has a Λ-isomorphism between

$$\Big(\bigoplus_{x\in\sigma} \varepsilon_x P_i \Big/ \bigoplus_{y\in\tau} \varepsilon_y P_i\Big)^{\#} \quad \text{and} \quad \bigoplus_{y\in\underline{\alpha}(\tau)} (\varepsilon_y P_i^{\#} \cap P_i^{\#}) \Big/ \bigoplus_{x\in\underline{\alpha}(\sigma)} (\varepsilon_x P_i^{\#} \cap P_i^{\#})$$

$$\Big(\cong \bigoplus_{y\in\underline{\alpha}(\tau)} (\varepsilon_y P_{\underline{\alpha}(i)} \cap P_{\underline{\alpha}(i)}) \Big/ \bigoplus_{x\in\underline{\alpha}(\sigma)} (\varepsilon_x P_{\underline{\alpha}(i)} \cap P_{\underline{\alpha}(i)})\Big) .$$

(ii) One has

$$\mathrm{Hom}_\Lambda\Big(P_j, \bigoplus_{x\in\sigma} \varepsilon_x P_i \Big/ \bigoplus_{y\in\tau} \varepsilon_y P_i\Big) \underset{R}{\cong} \mathrm{Hom}_\Lambda\Big(P_{\underline{\alpha}(i)}, \bigoplus_{x\in\underline{\alpha}(\sigma)} \varepsilon_x P_{\underline{\alpha}(j)} \Big/ \bigoplus_{y\in\underline{\alpha}(\tau)} \varepsilon_y P_{\underline{\alpha}(j)}\Big)$$

and

$$\mathrm{Hom}_\Lambda\Big(P_j, \bigoplus_{y\in\tau} (\varepsilon_y P_i \cap P_i) \Big/ \bigoplus_{x\in\sigma} (\varepsilon_x P_i \cap P_i)\Big) \cong$$

$$\underset{R}{\cong} \mathrm{Hom}_\Lambda\Big(P_{\underline{\alpha}(i)}, \bigoplus_{y\in\underline{\alpha}(\tau)} (\varepsilon_y P_{\underline{\alpha}(j)} \cap P_{\underline{\alpha}(j)}) \Big/ \bigoplus_{x\in\underline{\alpha}(\sigma)} (\varepsilon_x P_{\underline{\alpha}(j)} \cap P_{\underline{\alpha}(j)})\Big) .$$

Proof. (i) Let $i \in \{1,\dots,r\}$ be fixed and let $\delta\colon \mathfrak{Z}(KP_i) \to \mathfrak{Z}(KP_i^{\#})$ be the duality defined in (IV.9). In $\mathfrak{Z}(KP_i)$ one has the chain

$$(*)\qquad \bigoplus_{x\in\sigma} \varepsilon_x P_i \geq \bigoplus_{y\in\tau} \varepsilon_y P_i \geq P_i \geq \bigoplus_{y\in\tau} (\varepsilon_y P_i \cap P_i) \geq \bigoplus_{x\in\sigma} (\varepsilon_x P_i \cap P_i)$$

of Λ-lattices. Part (i) of (IV.10) is proved, if one shows that the chain

$$(*^{\#})\qquad \bigoplus_{x\in\underline{\alpha}(\sigma)} (\varepsilon_x P_i^{\#} \cap P_i^{\#}) \leq \bigoplus_{y\in\underline{\alpha}(\tau)} (\varepsilon_y P_i^{\#} \cap P_i^{\#}) \leq P_i^{\#} \leq \bigoplus_{y\in\underline{\alpha}(\tau)} \varepsilon_y P_i^{\#} \leq \bigoplus_{x\in\underline{\alpha}(\sigma)} \varepsilon_x P_i^{\#}$$

coincides with the chain obtained from (*) by applying δ. By (IV.9), $\delta(P_i) = P_i^{\#}$ and

$\delta(\bigoplus_{x\in\sigma} \varepsilon_x P_i) \cong (\bigoplus_{x\in\sigma} \varepsilon_x P_i)^{\#} \cong \bigoplus_{x\in\sigma} (\varepsilon_x P_i)^{\#} = \bigoplus_{x\in\sigma} \varepsilon_{\underline{\alpha}(x)} (\varepsilon_x P_i)^{\#}$. Hence,

$\delta(\bigoplus_{x\in\sigma} \varepsilon_x P_i) \subseteq \bigoplus_{x\in\underline{\alpha}(\sigma)} (\varepsilon_x P_i^{\#} \cap P_i^{\#})$, since $\bigoplus_{x\in\underline{\alpha}(\sigma)} (\varepsilon_x P_i^{\#} \cap P_i^{\#})$ is the biggest sublattice L of $P_i^{\#}$ with $\Lambda_{\underline{\alpha}(\sigma)} L = L$. But the smallest $L \in \mathfrak{Z}(KP_i)$ with $\Lambda_\sigma L = L$ and $P_i \leq L$ is $\bigoplus_{x\in\sigma} \varepsilon_x P_i$. Hence

$\delta(\bigoplus_{x\in\sigma} \varepsilon_x P_i) = \bigoplus_{x\in\underline{\alpha}(\sigma)} \delta\ (\varepsilon_x P_i^{\#} \cap P_i^{\#})$, since δ reverses inclusions and respects direct sums. That the other members of the chain $(*^{\#})$ coincide with the corresponding members of the chain $\delta(*)$ follows by the same arguments.

(ii) Choose primitive idempotents $\mathring{\varepsilon}_i \in \Lambda$ with $P_i \cong \Lambda\mathring{\varepsilon}_i$ for $1 \leq i \leq r$. One has the following chain of R-isomorphisms:

$$\mathrm{Hom}_\Lambda(P_j, \bigoplus_{x\in\sigma} \varepsilon_x P_i / \bigoplus_{y\in\tau} \varepsilon_y P_i) \cong \mathring{\varepsilon}_j (\bigoplus_{x\in\sigma} \varepsilon_x \Lambda\mathring{\varepsilon}_i / \bigoplus_{y\in\tau} \varepsilon_y \Lambda\mathring{\varepsilon}_i) \cong$$

$$\cong \bigoplus_{x\in\sigma} \varepsilon_x \mathring{\varepsilon}_j \Lambda \mathring{\varepsilon}_i / \bigoplus_{y\in\tau} \varepsilon_y \mathring{\varepsilon}_j \Lambda \mathring{\varepsilon}_i \cong \bigoplus_{x\in\underline{\alpha}(\sigma)} \varepsilon_x \alpha(\mathring{\varepsilon}_i)\Lambda\alpha(\mathring{\varepsilon}_j) / \bigoplus_{y\in\underline{\alpha}(\tau)} \varepsilon_y \alpha(\mathring{\varepsilon}_i)\Lambda\alpha(\mathring{\varepsilon}_j)$$

$$\cong \dots \cong \mathrm{Hom}_\Lambda(P_i^{\#}, \bigoplus_{x\in\underline{\alpha}(\sigma)} \varepsilon_x P_j^{\#} / \bigoplus_{y\in\underline{\alpha}(\tau)} \varepsilon_y P_j^{\#}) .$$

The proof of the second statement is similar.

q.e.d.

In terms of Theorem (IV.4)(v) and (vi) one gets from (IV.10)(i) $\lambda_{\underline{\alpha}(i)\underline{\alpha}(j)} = \mu_{ij}$. For instance, if Λ is a group ring, and all involved Brauer characters are real, then this means $\lambda_{ij} = \mu_{ij}$ and is a considerable refinement of (IV.4)(vi), which is often useful.

To proceed further with the general analysis of projective indecomposable lattices P_i a dual point of view of the inclusions in ( ) before (IV.2) is often helpful. There $\bigoplus_{s \in c_i} \varepsilon_s P_i$ was interpreted as the unique minimal superlattice of P_i which is completely decomposable. Dually, one can view P_i as a minimal sublattice L of $\bigoplus_{s \in c_i} \varepsilon_s P_i$ with $\varepsilon_s L = \varepsilon_s P_i$. Of course not every such sublattice is isomorphic to P_i . However, the following proposition leads to characterization of P_i as a sublattice of $\bigoplus_{s \in c_i} \varepsilon_s P_i$. This result does not depend on the basic assumptions of this chapter that Γ is a graduated order nor on the selfduality of Λ , but uses only the existence and properties of projective covers, pullbacks and pushouts. To get a general result one only has to modify the definition of c_i by $c_i = \{s \mid 1 \leq s \leq h, d_{si} \neq 0\}$.

(IV.11) Proposition. *Let x,y be nontrivial disjoint subsets of c_i for fixed i , $1 \leq i \leq r$.*

(i) There exists a Λ-module M which is an epimorphic image of both, $\varepsilon_x P_i$ and $\varepsilon_y P_i$, and satisfies the property: Each common epimorphic image of the two Λ-lattices $\varepsilon_x P_i$ and $\varepsilon_y P_i$ is epimorphic image of M . (Clearly, M is the amalgamating factor of $(\varepsilon_x + \varepsilon_y)P_i$ as amalgam of $\varepsilon_x P_i$ and $\varepsilon_y P_i$, i.e. $a_u^\sigma((\varepsilon_x+\varepsilon_y)P_i) \cong a_l^\sigma((\varepsilon_x+\varepsilon_y)P_i)$ with $\sigma = \{x, y, \{1,\dots,r\} \setminus (x \cup y)\}$, cf. (IV.3)).

(ii) For each subdirect product L of $\varepsilon_x P_i$ and $\varepsilon_y P_i$ (i.e. sublattice L of $\varepsilon_x P_i \oplus \varepsilon_y P_i$ with $\varepsilon_x L = \varepsilon_x P_i$ and $\varepsilon_y L = \varepsilon_y P_i$)

there exists an $\alpha \in Aut_\Lambda(\varepsilon_x P_i \oplus \varepsilon_y P_i)$ *with* $L\alpha \supseteq (\varepsilon_x+\varepsilon_y)P_i$.

Proof: Let $P = (\varepsilon_x+\varepsilon_y)P_i$. Then $\varepsilon_x P = \varepsilon_x P_i$ and $\varepsilon_y P = \varepsilon_y P_i$. Define $M := P/(\varepsilon_x P\cap P) \oplus (\varepsilon_y P\cap P)$ with $\nu : P \to M$ as the natural epimorphism, and define $\varphi_x : \varepsilon_x P \to M : \varepsilon_x 1 \to 1\nu$ and $\varphi_y : \varepsilon_y P \to M : \varepsilon_y 1 \to 1\nu$. Then φ_x , φ_y are well-defined Λ-epimorphisms onto M with kernels $\varepsilon_x P\cap P$ and $\varepsilon_y P\cap P$. Moreover $P = \{(a,b) \in \varepsilon_x P \oplus \varepsilon_y P \mid a\varphi_x = b\varphi_y\}$. Since each subdirect product L of $\varepsilon_x P_i$ and $\varepsilon_y P_i$ is of the form $L = \{(a,b) \in \varepsilon_x P \oplus \varepsilon_y P \mid a\tilde{\varphi}_x = b\tilde{\varphi}_y\}$ where $\tilde{\varphi}_x : \varepsilon_x P_i \to N$ and $\tilde{\varphi}_x : \varepsilon_y P_i \to N$ are epimorphisms of $\varepsilon_x P_i$ and $\varepsilon_y P_i$ onto a common epimorphic image N , it is clear that (ii) implies (i). Denote the epimorphism $L \to N : 1 \to (\varepsilon_x 1)\tilde{\varphi}_x = (\varepsilon_y 1)\tilde{\varphi}_y$ by ψ , with L , N etc. as just defined. Assume $N \neq 0$. Since $N/Jac(N) \cong S_i$ one has an epimorphism $\beta : P_i \to N$. Hence there exists a homomorphism $\gamma : P_i \to L$ with $\gamma\psi = \beta$. The first claim is that the image of γ is isomorphic to P . This follows because $P_i \cap (1-\varepsilon_x-\varepsilon_y)P_i \subseteq \ker\gamma$ and because the compositions $P_i \xrightarrow{\gamma} L \xrightarrow{\varepsilon_x} \varepsilon_x P_i$ and $P_i \xrightarrow{\gamma} L \xrightarrow{\varepsilon_y} \varepsilon_y P$ are epimorphisms (since $\varepsilon_x P_i$ and $\varepsilon_y P_i$ are local). Now (ii) follows easily if one observes that the A-module automorphism of $KL(= K\otimes_R L)$ induced by the Λ-isomorphism of $Im\gamma$ onto P maps $\varepsilon_x P$ and $\varepsilon_y P$ onto itself.

q.e.d.

For the (second part of the) next proposition the general assumptions of this chapter are needed again. The elements of the center $\Delta = Z(\Lambda)$ of Λ induce endomorphisms of the (projective indecomposable) Λ-lattices. One might ask how these endomorphisms behave with respect to the embeddings of the P_i into $\bigoplus_{s\in c_i} \varepsilon_s P_i$. Let $\omega_s : Z(A) \to Z(A_s) = Z_s$ denote the projection of the center of A onto Z_s , $1 \leq s \leq h$, cf. also end of Chapter III. One of the subtle points in the investigation of Λ (even if Γ is a graduated order is that $\omega_s(\Delta)$ is generally properly contained in the center $Z(\varepsilon_s\Lambda)$ of $\varepsilon_s\Lambda$. (See Theorem (VII.1)(i) for an example.)

(IV.12) Proposition. *Let s and t be different elements of c_i for some fixed i, $1 \leq i \leq r$, and let $\Delta = Z(\Lambda)$ be the center of Λ.*

(i) Define $I_{s,t} = \{x \in \omega_s(\Gamma) \mid x\varepsilon_s P_i \subseteq \varepsilon_s P_i \cap (\varepsilon_s + \varepsilon_t)P_i\}$ *and*

$I_{t,s} = \{x \in \omega_t(\Gamma) \mid x\varepsilon_t P_i \subseteq \varepsilon_t P_i \cap (\varepsilon_s + \varepsilon_t)P_i\}$.

Then there are homomorphisms $\nu_s : \omega_s(\Delta) \longrightarrow End_\Lambda(M_{s,t})$ *and* $\nu_t : \omega_t(\Delta) \longrightarrow End_\Lambda(M_{s,t})$ *with* $M_{s,t} = \varepsilon_s P_i \oplus \varepsilon_t P_i/(\varepsilon_s + \varepsilon_t)P_i$ *such that the kernel of* ν_s *is* $I_{s,t}$ *and the kernel of* ν_t *is* $I_{t,s}$ *and* $\nu_s(\omega_s(z)) = \nu_t(\omega_t(z))$ *holds for all* $z \in \Delta$. *In particular, if* $Z_s = K\varepsilon_s$ *and* $Z_t = K\varepsilon_t$ *one can identify* Z_s *and* Z_t *with* K *and obtains* $I_{s,t} = I_{t,s}$ *(as ideals of R) and* $\omega_s(z) \equiv \omega_t(z) \bmod I_{s,t}$ *for all* $z \in \Delta$.

(ii) If $c_i = \{s,t\}$, *then* $I_{s,t} = Z(\varepsilon_s\Lambda)\mathfrak{p}^{\tilde{\kappa}_s}$ *where* $\tilde{\kappa}_s$ *is the smallest integer bigger or equal than* $\frac{\kappa_s}{m_s}$ *with* κ_s *and* $\mathfrak{p}_s$ *as in (III.8).*

In particular, if $Z_s = K = Z_t$ *(as above) then* $u_s^{-1}R = u_t^{-1}R$ *and* $\omega_s(z) \equiv \omega_t(z) \bmod u_s^{-1}R$ *for all* $z \in \Gamma$.

Proof: (i) Since $M_{s,t}$ is the amalgamating factor of $(\varepsilon_s + \varepsilon_t)P_i$ as amalgam of $\varepsilon_s P_i$ and $\varepsilon_t P_i$ one has $M_{s,t} \cong \varepsilon_s P_i/\varepsilon_s P_i \cap (\varepsilon_s + \varepsilon_t(P_i \cong$
$\cong \varepsilon_t P_i/\varepsilon_t P_i \cap (\varepsilon_s + \varepsilon_t)P_i$.

Since $z \in \Delta$ operates on $\varepsilon_s P_i$ by multiplication with z as well as $\omega_s(z)$ and on $\varepsilon_t P_i$ by multiplication with z resp. $\omega_t(z)$, (i) follows.

(ii) In this case $\varepsilon_s P_i \cap (\varepsilon_s + \varepsilon_t)P_i = \varepsilon_s P_i \cap P_i$ and $\varepsilon_t P_i \cap (\varepsilon_s + \varepsilon_t)P_i = \varepsilon_t P_i \cap P_i$. The claim follows from part (i), (IV.2) and the conductor formula (III.8).

q.e.d.

The following application of the last two propositions can often be used for determining exponent matrices for the principal block of a group ring.

(IV.13) Corollary. *Let $\Lambda = RG$ be the group ring of a finite group G such that the characteristic $p(>0)$ of F does not divide the order $[G : G']$ of the commutator factor group $(char(K) = 0)$. Let S_1 be the trivial FG-module and V_1 the trivial KG-module. If the decomposition number d_{s1} for some fixed s, $1 < s \leq h$ is equal to 1 and $A_s \cong K^{l_s \times l_s}$, then*

$$m_{1i}^{(s)} + m_{i1}^{(s)} \leq a \quad \text{for at least one } i \in r_s, \ i \neq 1,$$

where a is the biggest natural number with $\omega_1(\bar{x}) \equiv \omega_s(\bar{x}) \bmod \mathfrak{p}^a$ for all class sums $\bar{x} \in Z(RG)$. (Note $\omega_s(\bar{x}) = \dfrac{|x|\chi_s(g)}{l_s}$ where χ_s is the character corresponding to V_s and $x = g^G$ the conjugacy class belonging to the class sum $\bar{x}$.)

Proof: Let L be the smallest sublattice of $\varepsilon_s P_1$ such that all composition factors of $\varepsilon_s P_1/L$ are isomorphic to S_1. The multiplicity of S_1 in the (unique) composition series of $\varepsilon_s P_1/L$ is equal to $b := \min\{m_{1i}^{(s)} \mid i \in r_s, i \neq 1\}$ (Note, by the normalization of the exponent matrix $m_{i1} = 0$ for $i \in r_s$.)

Since $[G : G']$ is relatively prime to p the module $\varepsilon_s P_1/L$ is centralized by G. Hence by (IV.11) $(\varepsilon_1 + \varepsilon_s)P_1$ has $\varepsilon_1 P_1/\mathfrak{p}^b \varepsilon_1 P_1 \cong \varepsilon_s P_1/L$ as amalgamating factor if viewed as amalgam of $\varepsilon_s P_1$ and the lattice $\varepsilon_1 P_1$ with the trivial G-action. From (IV.12) (i) one gets $\omega_1(\bar{x}) \equiv \omega_s(\bar{x}) \bmod \mathfrak{p}^b$. Hence $b \leq a$ and the inequality of (IV.13) follows.

q.e.d.

V. Blocks of multiplicity 1

The notation of the last chapter is kept. Assume that Λ is a block ideal of a group ring RG of a finite group G and that K is a splitting field of the block with minimal ramification (i.e. K is an unramified field extension of the p-adics $\mathbf{Q}_p$ to which values of the Frobenius characters in the block are adjoined). According to Michler [Mic 81] the multiplicity of the block Λ is defined by $\mu(\Lambda) := \max\{c_{ii} - 1 \mid i = 1,\dots,r\}$, where the c_{ii} are the diagonal entries of the Cartan matrix $C(\Lambda) = (c_{ij})_{1 \leq i,j \leq r}$ with $c_{ij} = \dim_F \mathrm{Hom}_{FG}(P_i/pP_i, P_j/pP_j)$.

Assume for the rest of this chapter that Λ has multiplicity 1 and denote the defect of Λ by d. Thus the diagonal Cartan numbers c_{ii} are equal to 2 and the decomposition numbers are equal to 0 or 1.

<u>*(V.1) Lemma.*</u> *K is unramified over $\mathbf{Q}_p$.*

Proof: Let $D = D(\Lambda) \in \mathbf{Z}_{\geq 0}^{h \times r}$ be the decomposition matrix of Λ in the usual sense without the modification of definition (III.10). If K is ramified over $\mathbf{Q}_p$ two irreducible Frobenius characters in the block Λ yield the same Brauer character by restriction to p-regular classes. Hence two rows of D must be equal. Since only two 1's can occur in each column $(C(\Lambda) = D^{tr}D$, $\mu(\Lambda) = 1)$, two columns of D are equal. This contradicts the fact that D has maximal rank r.

q.e.d.

Using the classification of root systems (known from the context of semisimple Lie algebras and Weyl groups) it is possible to describe the Cartan matrix up to a unimodular congruence transformation.

<u>*(V.2) Theorem.*</u> *Let* Λ *be a block of* RG *with multiplicity* 1 *. Then there exists a unimodular matrix* $g \in GL(r,\mathbb{Z})$ *with*

$$g^{tr}C(\Lambda)g = I_r + J_r$$

where I_r *,* $J_r \in \mathbb{Z}^{r\times r}$ *are the unit matrix and the matrix with all entries* 1 *resp.*

Proof: Since $C(\Lambda) = D^{tr}D$, the Cartan matrix $C(\Lambda)$ can be viewed as the matrix of the positive definite bilinear form induced by the standard scalar product of $\mathbb{Z}^{h\times 1}$ on the $\mathbb{Z}$-lattice L_D spanned by the columns of D . Since the vectors spanning L_D have square lengths 2, they satisfy the axioms of a root system and hence L_D is isometric to one of the so-called Witt lattices of type A_r, D_r, E_6, E_7, E_8, cf. e.g. [Hum 72], or an orthogonal direct sum of certain of these. But since Λ consists of one block only one sees that L_D cannot split into an orthogonal sum of proper sublattices (cf. [Kne 54] for a handy argument). Note, the determinant $\det C(\Lambda) = p^d$ is equal to the discriminant of L_D . The discriminants of E_6, E_7, E_8 are 3, 2, 1 resp. Hence L_D cannot be isometric to E_6, E_7, E_8 since there are less than 6 resp. 7 irreducible Brauer characters in a p-block of defect 1 for $p = 2$ resp. 3 and only one Brauer character for a block of defect zero. Moreover L_D cannot be isometric to D_r for $r \geq 4$. Assume they are isometric. D_r has a basis such that the matrix of scalar products is given by

$$\begin{pmatrix} 2 & 1 & & & & & & \\ 1 & 2 & 1 & & & 0 & & \\ & 1 & 2 & & & & & \\ & & & \ddots & & & & \\ & & & & 1 & & & \\ & & & 1 & 2 & 1 & 1 \\ & 0 & & & 1 & 2 & 0 \\ & & & & 1 & 0 & 2 \end{pmatrix} .$$

To obtain a basis $x_1,\dots,x_r$ of $L_D (\subseteq \mathbb{Z}^{h\times 1})$ with the same matrix of scalar products one has to choose

$x_1 = e_1 + e_2$, $x_2 = e_2 + e_3, \dots, x_{r-1} = e_{r-1} + e_r$, $x_r = e_{r-1} - e_r$ after a suitable permutation with sign changes of the standard basis $e_1,\dots,e_h$ of $\mathbb{Z}^{h\times 1}$. Since no row of the decomposition matrix D is zero, none of the vectors $e_1,\dots,e_h$ can be orthogonal to L_D, one concludes $h = r$. But a block with the same number of irreducible Brauer- and Frobenius characters is of defect zero, hence $r = h = 1$, contradicting $r \geq 4$. Since D_3 is isometric to A_3 the lattice L_D is isometric to A_r, which can be realized, e.g. as $A_r = \{(a_0,\dots,a_t)^{tr} \in \mathbb{Z}^{(r+1)\times 1} \mid a_0 + \dots + a_r = 0\}$ with the bilinear form induced from the standard scalar product of $\mathbb{Z}^{(r+1)\times 1}$. Choosing $e_1 - e_0$, $e_2 - e_0, \dots, e_r - e_0$ as basis for A_r $(e_0,\dots,e_r)$ is the standard orthogonal basis of $\mathbb{Z}^{(r+1)\times 1}$), one obtains $I_r + J_r$ as matrix of the scalar product.

q.e.d.

(V.3) Corollary [Mic 81]: Let Λ be as in (V.2). Then $\det C(\Lambda) = p^d = r + 1 = h$, where r and h are the numbers of irreducible Brauer resp. Frobenius characters in the block Λ. Furthermore $C(\Lambda)$ has one elementary divisor equal to p^d and all the others equal to 1.

Proof: $\det C(\Lambda) = \det(I_r + J_r) = r + 1$ by (V.2) (note, the characteristic polynomial of J_r is $(-x)^{r-1}(r-x)$, inserting $x=-1$ yields the determinant). Hence $r+1$ is a power of p by Brauer's result on the determinant of the Cartan matrix. But one easily produces matrices $g,h \in GL(r,\mathbb{Z})$ with $g(I_r+J_r)h = \mathrm{diag}(1,\dots,1,r+1)$. Hence by Brauer's determination of the elementary divisors of the Cartan matrix of a block (cf. [Bra 56], §5,6) one gets $r+1=p^d$, where d is the defect of Λ . Finally $r+1=h$ follows independently from the preceding results (cf. [Mic 81] 2.1.c): Clearly $h>r$, since Λ is not of defect zero. On the other hand, if one constructs the decomposition matrix D columnwise in such a way that each new column d_i is not orthogonal to all the preceding ones $d_1,\dots,d_{i-1}$, at most $i+1$ rows of the matrix with columns $d_1,\dots,d_i$ are not equal to zero, since there are exactly two ones in each column $(1 \le i \le r)$. Hence $h \le r+1$, which implies $h=r+1$.

q.e.d.

The further investigation of Λ will not depend on (V.2) and (V.3). Let the sets r_s, c_i , the exponent matrices $M^{(s)} = (m_{ij}^{(s)})$ $(1 \le s \le h$, $1 \le i \le r)$ etc. be defined as at the beginning of Chapter IV. Since Λ is of multiplicity one, $|c_i| = c_{ii} = 2$ for $i=1,\dots,r$. Hence the amalgamation matrix $\alpha(P_i)$ of P_i consists of only two rows (cf. (IV.6)). Moreover, for $i \neq j$ $(1 \le i,j \le r)$, one has $c_{ij} = |c_i \cap c_j|$ equal to 0 or 1 . The results of the last chapter yield the following theorem without much effort.

__(V.4) Theorem.__ Let Λ be a block of RG with multiplicity one and defect d .

(i) [Mic 81] All irreducible Frobenius characters χ_s $1 \le s \le h$ in Λ have height zero, i.e. p^d is the highest power of p dividing

$\frac{|G|}{\chi_s(1)}$ *for each index* $1 \leq s \leq h$.

(ii) *The central characters (given by* $\omega_s(g) = \frac{|g^G|\chi_s(g)}{\chi_s(1)}$ *for* $g \in G$*) satisfy the congruences* $\omega_s(g) \equiv \omega_t(g) \pmod{p^d R}$ *for all* $g \in G$.

(iii) *The exponent matrices* $M^{(s)} = (m_{ij}^{(s)})_{i,j \in r_s}$ *of* $\varepsilon_s\Lambda$ *satisfy* $m_{ij}^{(s)} + m_{ji}^{(s)} = d$ ($= \varkappa_s$ *cf.* (III.8)) *for* $i \neq j \in r_s$, $s = 1,\ldots,h$.

(iv) *For each* i , $1 \leq i \leq r$ *the upper and lower amalgamating factors* $a_u(P_i)$ *and* $a_l(P_i)$ *are isomorphic. They are uniserial with* $S_i = P_i/Jac(P_i)$ *as only composition factor, the multiplicity of* S_i *in* $a_u(P_i)$ *is* d , *and the annihilator* $\{x \in R \mid xa_u(P_i) = 0\}$ *of* $a_u(P_i)$ *in* R *is equal to* $p^d R$.

(v) *The conductor of* $\bigoplus_{s=1}^{h} \varepsilon_s\Lambda$ *in* Λ *is* $\bigoplus_{s=1}^{h} \Lambda'_s$ *where* $\Lambda'_s \subseteq \varepsilon_s\Lambda$ *has the exponent matrix* $M^{(s)} - dI_{|r_s|}$ *(here* I_n *denotes the* $n\times n$*-unit matrix).*

(vi) *If all Frobenius characters (and hence all Brauer characters) in the block* Λ *are real, the exponent matrices* $M^{(s)}$ *are of two possible kinds:*
$M^{(s)} = \begin{pmatrix} 0 & d \\ 0 & 0 \end{pmatrix}$ *or* $M^{(s)} = \frac{d}{2}(J_{|r_s|} - I_{|r_s|})$, $(1 \leq s \leq h)$, *where* J_n *denotes the* $n\times n$*-matrix with all entries equal to* 1 . *(Of course* d *is even, if for one* s *with* $|r_s| > 1$, *the last case occurs.)*

Proof: (i)-(v) follow immediately by looking at the amalgamation matrices $\alpha(P_i)$ of the projective indecomposables $P_1,\ldots,P_r$. Fix an i , $1 \leq i \leq r$. Then $\alpha(P_i)$ has two rows with indices, say, s and t , i.e. $c_i = \{s,t\}$. Now the claim is $r_s \cap r_t = \{i\}$. Clearly $i \in r_s \cap r_t$. On the other hand, if $j \in r_s \cap r_t$, $j \neq i$, then the i-th and the j-th column of the decomposition matrix D must be equal, namely 1's in the positions s and t and 0's elsewhere, since the multiplicity of the block is 1 . But this is a contradiction, since D has maximum rank. Hence $r_s \cap r_t = \{i\}$, and after rearranging the column indices $1,\ldots,r$ of $\alpha(P_i)$, the amalgamation matrix looks as follows:

$$\alpha(P_i):\qquad \begin{array}{c|cccc} & \overbrace{*\ldots*\ \ \varkappa_s}^{r_s} & & & \\ & & \overbrace{\varkappa_t\ \ *\ldots*}^{r_t} & & \\ & & i & & \\ \hline s & *\ldots* & \varkappa_s & 0\ldots0 & 0\ldots0 \\ t & 0\ldots0 & \varkappa_t & *\ldots* & 0\ldots0 \end{array}$$

where the stars indicate the entries $\varkappa_s - m_{ij}^{(s)} - m_{ji}^{(s)}$ for $i,j \in r_s$, $i \neq j$, resp. $\varkappa_t - m_{ij}^{(t)} - m_{ji}^{(t)}$ for $i,j \in r_t$, $i \neq j$. Moreover, since K is unramified over $\mathbb{Q}_p$, $\varkappa_s$ is the biggest natural number with $p^{\varkappa_s} \mid \frac{|G|}{\chi_s(1)}$, analogously $\varkappa_t$. Since $\alpha(P_i)$ has only two rows, the entries in the same column must be equal by (IV.7).

By the connectedness of the Brauer graph all $\varkappa_u$ must be equal for $1 \leq u \leq h$, and hence all equal to d, by the definition of the defect. This proves (i). (iii) follows from this by the above equations (cf. also (IV.7) (i)), (iv) and (v) now follow (IV.4), (IV.5), and (III.8), and (ii) follows from (IV.12). Finally (v) follows from (IV.1), more precisely from the equation for $m_{ij}^{(s)}$ right after (IV.1) (for real Brauer and Frobenius characters). Note for the application of this equation that all entries in the first row of $M^{(s)}$ except for the first position are equal to d, if one chooses the first column of $M^{(s)}$ to be zero.

q.e.d.

The situation of (V.4) is favourable in the following respect: Once the exponent matrices $M^{(s)}$ are known, one has a complete description of the isomorphism type of the ring Λ. Namely Λ is isomorphic to the subring of all matrix tuples $((a_{ij}^{(s)})_{i,j\in r_s})_{s=1,\ldots,h}$ of $\bigoplus_{s=1}^{h} \Lambda(\tilde{n}^{(s)}, M^{(s)})$ satisfying $a_{ii}^{(s)} \equiv a_{ii}^{(t)} \pmod{\mathfrak{p}^d}$ whenever $c_i = \{s,t\}$.

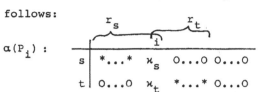

(Note, $a_{ij}^{(s)} \in R^{n_i \times n_j}$ with $n_i = \dim_F S_i$, $n_j = \dim_F S_j$, $i,j \in r_s$, $s = 1,\dots,h$.)

However, the exponent matrices are not completely determined except in the situation (V.4)(vi), typical examples of which are the blocks of defect 1 of symmetric groups or the nonprincipal 2-block of $PSU_3(5)$ of defect 2 at the prime $p = 2$. In the first case the block Λ is Morita equivalent to

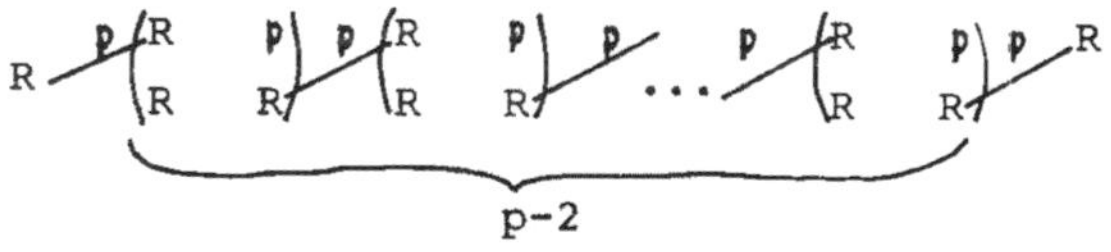

where $\mathfrak{p}^i$ indicates a congruence modulo $\mathfrak{p}^i$. Of course, this follows already from Brauer's classical defect-1-theory [Bra 41]. In the second example (of $G = PSU_3(5)$) the decomposition numbers are given by

	28_1	28_2	28_3
28_1	1	0	0
28_2	0	1	0
28_3	0	0	1
84	1	1	1

(here and in the sequel characters are denoted by their degrees). Hence by (V.4)(vi) Λ is Morita equivalent to

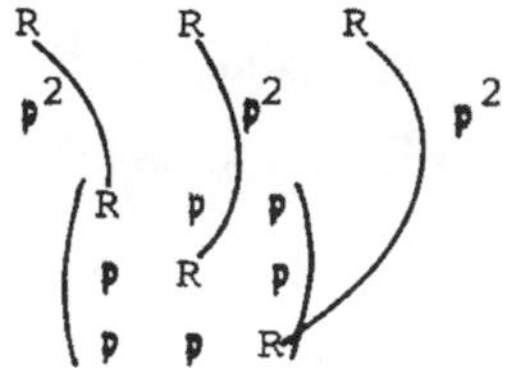

$(\subseteq R \oplus R \oplus R \oplus R^{3\times 3})$.

The decomposition matrix $D_{p^d} = \begin{pmatrix} I_{p^d-1} \\ 1\dots1 \end{pmatrix} \in \mathbf{Z}^{p^d \times (p^d-1)}$ seems to occur quite often for blocks of multiplicity one namely in case G is p-solvable by the Fong-Swan Theorem and also if Λ is the principal 2-block (of multiplicity 1) of a finite group with abelian 2-Sylow subgroup, as Michler proved, cf. [Mic 81], Thm.5.2. As proved there as well, the only simple groups with principal 2-block of multiplicity one are the $PSL_2(q)$, $q \equiv 3 \pmod 8$, $q > 3$.

(V.5) Example: *Let* Λ *be the principal 2-block of* RG *with* $G = PSL_2(q)$, $q \equiv 3 \pmod 8$. *The decomposition numbers are given by* D_4 , *and the graduated hull of* Λ *by*

$$\Lambda(1;(0)) \oplus \Lambda(\tfrac{q-1}{2};(0)) \oplus \Lambda(\tfrac{q-1}{2};(0)) \oplus \Lambda(1,\tfrac{q-1}{2},\tfrac{q-1}{2};\begin{pmatrix}0&2&2\\0&0&1\\0&1&0\end{pmatrix})$$

Proof: By Brauer [Bra 66] there are three irreducible Brauer characters in the block Λ , which is of defect 2. They are obtained by restricting the trivial character 1 and the two complex conjugate characters of degree $\frac{q-1}{2}$ to the 2-regular classes. This yields the decomposition matrix. Only the exponent matrix of the graduated order corresponding to the irreducible Frobenius character has to be determined, since the other Frobenius characters yield maximal orders. The first row and column of this exponent matrix are determined by (V.4)(iii). That the two missing non-diagonal entries are equal and hence equal to 1, follows from (IV.1)(i) by using an outer automorphism of $PSL_2(q)$ of order 2 interchanging the two characters of degree $\frac{q-1}{2}$. (For instance the automorphism induced by $g \to g^{-tr}$ of $SL_2(q)$ will do.)

q.e.d.

For a good portion of the preceding arguments it was not really used that Λ is a block in a group ring RG . Hence there are generalizations to selfdual (2-sided indecomposable) orders with decomposition matrix of maximal rank and diagonal Cartan numbers $c_{ii} = 2$. Here is an application of these obvious generalizations.

<u>*(V.6) Example:*</u> *Let* Λ' *be principal 2-block of* RG *with* $G = SL_2(q)$, $q \equiv 3 \pmod 8$, *let* $Z(G) = \{1, z\}$ *be the center of* G , *and* $E = \frac{1}{2}(1-z) \in KG$ *be the sum of the central primitive idempotents corresponding to the faithful irreducible Frobenius characters of* G . *Define* $\Lambda = E\Lambda'$. *(Thus* Λ *can be viewed as block of a twisted group ring of* $PSL_2(q)$, *cf. (III.2), (III.4). The decomposition numbers of* Λ *are given by*

	1	$\frac{q-1}{2}$	$\overline{\frac{q-1}{2}}$
$\frac{q+1}{2}$	1	1	0
$\overline{\frac{q+1}{2}}$	1	0	1
$q-1$	0	1	1

and all three exponent matrices are given by $\begin{pmatrix} 0 & 1 \\ 0 & 0 \end{pmatrix}$. *Moreover* Λ *is Morita equivalent to* $\mathfrak{p}\begin{pmatrix} R & \mathfrak{p} \\ R & R \end{pmatrix}\mathfrak{p}\begin{pmatrix} R & \mathfrak{p} \\ R & R \end{pmatrix}\mathfrak{p}\begin{pmatrix} R & \mathfrak{p} \\ R & R \end{pmatrix}$ $\left(\subseteq \bigoplus_{i=1}^{3} \begin{pmatrix} R & \mathfrak{p} \\ R & R \end{pmatrix}\right)$

Proof: The irreducible Brauer characters of Λ' can be identified with those in (V.5). This yields the decomposition matrix of Λ . The exponent matrices follow from the proper generalization of (V.4) (iii), where one has to keep in mind that d has to be chosen equal to 1 , namely compare proof of (III.2) and keep in mind that 2 still divides the character degrees $\frac{q+1}{2}$ and $q-1$.

q.e.d.

It might be worthwhile to mention that the two orders denoted by Λ in (V.5) and (V.6) have the same discriminant (with respect to the regular trace) and, if taken modulo 2, yield isomorphic F-algebras b, namely the principal block of $F\,PSL_2(q)$. Indeed, the principal block Λ' of $R\,PSL_2(q)$ is isomorphic to the pullback of the diagram of the epimorphisms of the two orders Λ' and Λ onto b .

As a final example a somewhat bigger exponent matrix will be discussed.

(V.7) Example: *Let G be the 2-transitive Frobenius group $Aff(1,\mathbb{F}_9) = \{\begin{pmatrix} a & b \\ 0 & 1\end{pmatrix} \mid a,b \in \mathbb{F}_9,\ a \neq 0\} < GL_2(9)$ of order $9 \cdot 8$ and let $p=3$. G has one irreducible Frobenius character χ of degree 8 and 8 one dimensional characters $\chi_0, \chi_1, \ldots, \chi_7$ numbered in such a way that χ_i is the i-th power of χ_1 . $\Lambda = RG$ consists of one block with multiplicity 1 and defect 2 , the irreducible Brauer characters are the restrictions $\hat{\chi}_0, \ldots, \hat{\chi}_7$ of the χ_i to 3-regular classes, the decomposition matrix of Λ is $D_8 (= \begin{pmatrix} & I_8 & \\ 1 & \ldots & 1\end{pmatrix})$ and $\varepsilon_\chi \Lambda = \Lambda(1,1,\ldots,1,M)$ with exponent matrix*

$$M = \left(\begin{array}{c|cc|cc:c|cc} 0 & 2 & 2 & 2 & 2 & 2 & 2 & 2 \\ \hline 0 & 0 & 1 & 2 & 1 & 2 & 2 & 2 \\ 0 & 1 & 0 & 1 & 2 & 2 & 2 & 2 \\ \hline 0 & 0 & 1 & 0 & 1 & 1 & 2 & 1 \\ 0 & 1 & 0 & 1 & 0 & 1 & 1 & 2 \\ \hdashline 0 & 0 & 0 & 1 & 1 & 0 & 2 & 2 \\ \hline 0 & 0 & 0 & 0 & 1 & 0 & 0 & 1 \\ 0 & 0 & 0 & 1 & 0 & 0 & 1 & 0 \end{array}\right)$$

where the Brauer characters are arranged in the above order. (Note, the ring structure of RG is completely determined by this information.)

Proof: After choosing the first column of M to be equal to zero, (V.4) (iii) implies that the nondiagonal elements of the first row are equal to 2. It is convenient to number the simple FG-modules as well as the irreducible Brauer characters by the integers mod 8.

Then the Frobenius automorphism of $\mathbb{F}_9$ induces an automorphism of order 2 of G which permutes the $\hat{\chi}_i$ by $\hat{\chi}_i \to \hat{\chi}_{3i}$ $(i \in \mathbf{Z}/8\mathbf{Z})$. The standard antiautomorphism $g \to g^{-1}$ of G induces the permutation $i \to -i$ of the index set $\mathbf{Z}/8\mathbf{Z}$. Finally the automorphism of RG induced by the linear character χ_j (cf. automorphism type (ii) introduced before (IV.1)) induces the permutation $i \to i+j$ of the indices of the matrix $M = (m_{ij})_{i,j \in \mathbf{Z}/8\mathbf{Z}}$. The structural invariants $m_{ijk} = m_{ij} + m_{jk} + m_{ik}$ satisfy $(m_{ij} =) m_{ij0} = m_{3i\ 3j\ 0} = m_{0\ -j-i} = m_{i\ i-j\ 0} =$

$= m_{3i\ 3(i-j)\ 0}$; hence $m_{ij} = m_{3i\ 3j} = m_{i\ i-j} = m_{3i\ 3(i-j)}$ for $i,j \in \mathbf{Z}/8\mathbf{Z}$. If one sets $a = m_{13}$, $b = m_{12}$, $c = m_{14}$, $d = m_{26}$, $e = m_{25}$, $f = m_{27}$, $g = m_{45}$, $h = m_{57}$, the following part of M is determined by these equations alone:

$$M = \begin{array}{c|cc|cc|c|cc}
0 & 2 & 2 & 2 & 2 & 2 & 2 & 2 \\ \hline
0 & 0 & a & b & a & c & c & b \\
0 & a & 0 & a & b & c & b & c \\ \hline
0 & & & 0 & d & d & e & f \\
0 & & & d & 0 & d & f & e \\ \hdashline
0 & & & & & 0 & g & g \\ \hline
0 & & & & h & & 0 & h \\
0 & & & h & & & h & 0
\end{array}$$

(Note the indices come in the order 0,1,3,2,6,4,5,7.)

Applying (V.4)(iii) one obtains (in the notation $x' = 2-x$ for $x \in \mathbf{Z}$): $1 = a = a' = d = d' = h = h' = f = f'$ and M is determined if b, c, e, and g are known.

Let $\overline{P}_0 = P_0/\mathfrak{p}P_0$ be the projective indecomposable FG-module with top S_0. It is an easy corollary of Jenning's Theorem (cf. [HuB82]) that the Loewy series of $\overline{P}_0$ is given by

$$\begin{array}{ccccc}
 & & S_0 & & \\
 & S_1 & & S_3 & \\
S_2 & & S_4 & & S_6 \\
 & S_5 & & S_7 & \\
 & & S_0 & &
\end{array}$$

Hence one obtains the Loewy series of $L/\mathfrak{p}L$ with $L = e_\chi P_0$ by deleting the socle S_0. This immediately implies that b, c, e, g are bigger than zero. Moreover sublattices of L of the form $L(\tilde{m})$ where $\tilde{m} = (1,0,0,a_2,a_6,a_4,a_5,a_7)^{tr}$ with $a_i \in \mathbb{Z}_{\geq 0}$ and $a_i > 0$ for at least one $i \in \{2,6,4,5,7\}$ do not exist. This implies $b' = c' = e' = g' = 0$ and the claim follows.

q.e.d.

In [Ple 80a] the group ring $G = \mathrm{Aff}(1,\mathbb{F}_8)$ at the prime 2 (defect 3) is treated in the same way (without applying Jenning's Theorem).

VI. Examples of group rings

In this chapter the group rings of some individual groups are discussed to demonstrate how the results of the Chapters II, III, and mainly IV can be applied to compute the graduated hull of a group ring with decomposition numbers 0 and 1. The basic notation of Chapter IV is kept, and p denotes the characteristic of $F = R/\mathfrak{p}$ and K is a finite extension field of $\mathbb{Q}_p$. The following result is often a good starting point for groups with a 2-transitive permutation representation.

(VI.1) Theorem (cf. [Ple 77] Thm. 5.1 for the global version). *Let G act 2-transitively on the set N, denote by RN the RG-permutation lattice, and let $L = RN/R\sum_{w\in N} w$ the non trivial irreducible factor lattice.*

If the characteristic p of $F = R/\mathfrak{p}$ divides $|N|$ and $L/\mathfrak{p}L$ has exactly two (FG-)composition factors (one of which is the trivial FG-module), then all RG-sublattices of L are of the form $\mathfrak{p}^{\beta}L_{(\alpha)}$ with $L_{(\alpha)} = \{\sum_{w\in N} x_w\bar{w} \mid x_w \in R,\ \sum_{w\in N} x_w \in \mathfrak{p}^{\alpha}\}$, where $\beta \in \mathbb{Z}_{\geq 0}$, $\alpha = 0,1,\ldots,t$ with $\mathfrak{p}^t = |N|R$ and where $\bar{w}$ denotes the image of $w \in N \subseteq RN$ in L.

If moreover the nontrivial constituent of $L/\mathfrak{p}L$ is absolutely irreducible, then $\varepsilon RG \cong \Lambda(1, |N|-2, \begin{pmatrix} 0 & t \\ 0 & 0\end{pmatrix})$ with t as defined above and ε the central primitive idempotent of KG with $\varepsilon L = L$.

If one assumes only that the trivial FG-constituent occurs only once in $L/\mathfrak{p}L$, then $L/L_{(t)}$ is still the biggest factor module of L, on which G acts trivially.

Proof: The first part can be proved by the same argument as the G-lattice version in [Ple 77]. The second statement follows from the first part, by (II.8) and (II.4). The third statement follows in the same way as the first.

q.e.d.

Examples for permutation groups satisfying the hypotheses of (VI.1) are symmetric and alternating groups S_n and A_n with the natural permutation representation with the exception of A_4, $p=2$ (cf. [Far 62] for S_n and [Kle 75] for other examples). In the case of the symmetric group (VI.1) can be extended to the exterior powers of L. For any RG-module M, let $\Lambda^k(M)$ denote the k-th exterior power of M, $\Lambda^k(M) = \bigotimes_{i=1}^{k} M / \langle m_1 \otimes \ldots \otimes m_k \mid m_1,\ldots,m_k \in M$ with $m_i = m_j$ for two indices i,j, $1 \leq i < j \leq k \rangle$.
Clearly $\Lambda^k(M)$ is an RG-module of R-rank $\binom{m}{k}$ where m is the R-rank of M. If M is a lattice or an FG-module, the same holds for $\Lambda^k(M)$. The character value for $g \in G$ on $\Lambda^k(M)$ is given by the k-th elementary symmetric function of the eigenvalues of the action of g on M (in a suitable field) $0 \leq k \leq m$. With this in mind, one checks (cf. [Ple 80a] III.7) that the Frobenius character for $\Lambda^k(L)$, where L is the natural RS_n-lattice described in (VI.1) (for $|N| = n$), is denoted by $[n-k, 1^k]$ in the usual notation for characters of symmetric groups, cf. [JaK 81].

__(VI.2) Theorem__ ([Ple 80a] III.8). Let $p = char(F)$ be an odd divisor of n, G the symmetric group S_n of the finite set $\{1,2,\ldots,n\}$, and let $L, L_{(\alpha)}$ be defined as in (VI.1). Then for $1 \leq k < n$ the RG-sublattices of $\Lambda^k(L)$ are given by $p^{\beta}\Lambda^k(L_{(\alpha)})$ for $\beta \in \mathbb{Z}_{\geq 0}$, $\alpha = 0,1,\ldots,t$ where $p^t = Rn$. If $\varepsilon_{(k)}$ is the central primitive idempotent of KG with $\varepsilon_{(k)}\Lambda^k(L) = \Lambda^k(L)$ then $\varepsilon_{(k)}RG \cong \Lambda(\binom{n-2}{k-1}, \binom{n-2}{k}, \begin{pmatrix} 0 & t \\ 0 & 0 \end{pmatrix})$.

Proof: By a result by Peel [Pee 71], $\Lambda^k(L/\mathfrak{p}L)$ has two nonisomorphic irreducible constituents (cf. the remarks on the identification of the character of $\Lambda^k(L)$ preceding this theorem). These constituents are $\Lambda^{k-1}(L_{(1)}/\mathfrak{p}L)$ and $\Lambda^k(L_{(1)}/\mathfrak{p}L)$, are absolutely irreducible and have F-dimensions $\binom{n-2}{k-1}$ and $\binom{n-2}{k}$. To prove the theorem it therefore suffices to check that $\Lambda^k(L/\mathfrak{p}L) \cong \Lambda^k(L)/\mathfrak{p}\Lambda^k(L)$ is indecomposable (because then by the same duality argument as in the proof of (VI.1) $\Lambda^k(L_{(t)},\mathfrak{p}L_{(t)}) \cong \Lambda^k(L_{(t)})/\mathfrak{p}\Lambda^k(L_{(t)})$ is indecomposable). By Peel's theorem quoted above, $\Lambda^{k-1}(L_{(1)}/\mathfrak{p}L)$ and $\Lambda^k(L_{(1)}/\mathfrak{p}L)$ remain irreducible and nonisomorphic if restricted to the stabilizer $\mathrm{Stab}_G(i)$ of $i = 1,2,\ldots,$ or n, i.e. to S_{n-1}. As $R\mathrm{Stab}_G(i)$-module $L/\mathfrak{p}L$ decomposes into a direct sum of two irreducible nonisomorphic $R\mathrm{Stab}_G(i)$-modules such that the $\mathrm{Stab}_G(i)$-complements, $i = 1,\ldots,n$, of $L_{(1)}/\mathfrak{p}L$ are permuted transitively by S_n. Hence the same holds for $\Lambda^k(L/\mathfrak{p}L)$, and the result follows.

q.e.d.

In the sequel characters will be denoted by their degrees. As an easy example to start with the symmetric group S_{10} on ten symbols will be discussed at the prime 5 .

(VI.3) Example Let $G = S_{10}$ be the symmetric group of degree 10, R the 5-adic integers (or the localization of $\mathbb{Z}$ at (5) can be chosen as R as well). Let Λ be the principal block ideal of RG (the other blocks have defect 0 and 1). Then the decomposition numbers are

	1	8	28	56	1'	8'	28'	56'	70	34	217	34'	217'	266
1	1													
9	1	1												
36		1	1											
84			1	1										
1'					1									
9'					1	1								
36'						1	1							
84'							1	1						
126				1					1					
126'								1	1					
42		1								1				
288	1	1	1							1	1			
42'						1						1		
288'					1	1	1					1	1	
567			1	1							1			1
567'							1	1					1	1
252	1										1	1		
252'					1					1			1	
448				1				1	1					1
768										1	1	1	1	1

For $s = 1, 1'$ one has $\varepsilon_s\Lambda \cong R$,
for $s = 9, 36, 84, 9', 36', 84', 126, 126', 42,$ and $42'$ the exponent matrix of $\varepsilon_s\Lambda$ is $\begin{pmatrix} 0 & 1 \\ 0 & 0 \end{pmatrix}$. The other exponent matrices of $\varepsilon_s\Lambda$ (with the Brauer characters ordered as in the decomposition matrix above) are:

$$M = \begin{pmatrix} 0 & 1 & 2 & 2 & 1 \\ 0 & 0 & 1 & 1 & 1 \\ 0 & 0 & 0 & 1 & 0 \\ 0 & 0 & 1 & 0 & 0 \\ 0 & 1 & 1 & 1 & 0 \end{pmatrix} \qquad \text{for } s = 288, 288'$$

(i.e. $e_{288}\Lambda = \Lambda(1,8,28,34,217,M)$),

$$\begin{pmatrix} 0 & 1 & 1 & 2 \\ 0 & 0 & 1 & 1 \\ 0 & 1 & 0 & 1 \\ 0 & 0 & 0 & 0 \end{pmatrix}$$ *for* $s = 567, 567'$,

$$\begin{pmatrix} 0 & 1 & 2 \\ 0 & 0 & 1 \\ 0 & 0 & 0 \end{pmatrix}$$ *for* $s = 252, 252'$, *(for* $s = 252'$ *reorder the Brauer characters :* $1', 217', 34$*)*

$$\begin{pmatrix} 0 & 2 & 1 & 1 \\ 0 & 0 & 0 & 0 \\ 0 & 1 & 0 & 1 \\ 0 & 1 & 1 & 0 \end{pmatrix}$$ *for* $s = 448$, *and* $$\begin{pmatrix} 0 & 1 & 2 & 1 & 2 \\ 0 & 0 & 1 & 1 & 1 \\ 0 & 0 & 0 & 0 & 1 \\ 0 & 1 & 1 & 0 & 1 \\ 0 & 0 & 1 & 0 & 0 \end{pmatrix}$$ *for* $s = 768$.

Proof: Since K is a splitting field of S_{10} , the hypotheses of Chapter IV are satisfied. From the decomposition number, cf. [Mac 76], one has $\varepsilon_s\Lambda \cong R$ for $s = 1, 1'$, and by (VI.2) $\varepsilon_s\Lambda$ has exponent matrix $\begin{pmatrix} 0 & 1 \\ 0 & 0 \end{pmatrix}$ for $s = 9, 36, 84, 9', 36', 84', 126, 126'$. Since the defect is 2 the only possible entries for the exponent matrices are $0, 1$, and 2 . The exponent matrices are normalized in such a way that the first column is zero. Let $N \in \mathbb{Z}_{\geq 0}^{r \times r}$ with $r > 2$ be an exponent matrix of some $\varepsilon_s\Lambda$. The basic observation is that N is completely determined by its first row $(0, n_{12}, \ldots, n_{1k})$. Namely let n_{ij} $(2 \leq i, j \leq k$, $i \neq j)$ any other off diagonal entry of N , then $n_{ij} + n_{ji} = 1$ or $= 2$ by (II.2)(*) and (III.8) (defect 2), and $n_{ij} - n_{ji} = n_{1j} - n_{1i}$ since all characters are real, cf. (IV.1). Taking the second equation modulo 2 , yields a decision which possibility holds for the first equation. After this one can find the unique solution of these linear equations. Conversely, if $n_{ij} + n_{ji} = 2$ is known for some $2 \leq i, j \leq k$, $i \neq j$, then the two equations imply $n_{1i} = n_{1j}$.

To get this starting information, observe that $c_i \cap c_j$ contains either $0, 1$, or 2 elements for $i \neq j$, i, j any irreducible Brauer character of Λ (look at the Cartan matrix !). If $c_i \cap c_j = \{s\}$ one applies (IV.7) (i) to obtain $m_{ij}^{(s)} + m_{ji}^{(s)} = 2$, and if $c_i \cap c_j = \{s, t\}$

with $s \in \{9,36,84,9',36',84',126,126'\}$ and t not in this set, one applies (IV.7) (ii) and the beginning of the proof to obtain $m_{ij}^{(t)} + m_{ji}^{(t)} = 1$. This information together with the above remarks suffices to determine all exponent matrices with more than two rows.

For example, the first row of the exponent matrix of $\varepsilon_{288}\Lambda$ must be $(0,1,2,2,x)$ with $x = 1$ or $x = 2$ by these arguments. By the same argument $m_{8,217}^{(288)} + m_{217,8}^{(288)} = 2$, hence $x = 1$. This determines the exponent matrix M as given in the formulation of the example by the equation derived above. Proceeding in this way one can determine the other exponent matrices in the same order as they are listed; (note, one sometimes has to use the previous exponent matrices for the determination of the later ones). Finally the exponent matrix of $\varepsilon_s\Lambda$, $s = 42,42'$ follows immediately from that of $\varepsilon_{288}\Lambda$ resp. $\varepsilon_{288'}\Lambda$.

q.e.d.

By essentially the same method S_6, S_7, S_8 are treated at the prime 3 in [Ple 80a], as well as the principal block of S_7 at the prime 2 , which is somewhat more difficult. In the next example the defect is 4 and the character values generate a ramified extension field of the ground field.

(VI.4) Example. Let $G = PSL_3(3)$, R the ring of 2-adic integers, and Λ the principal block ideal of RG . The decomposition numbers in the modified form as defined in (III.10) are given by

	1	12	26
1	1		
12		1	
13	1	1	
26			1
26'			1
27	1		1
39	1	1	1

where two absolutely irreducible Frobenius characters belong to 26' *, more precisely the center of* $\varepsilon_{26'}A$ *is isomorphic to* $K[\sqrt{-2}]$.
One has $\varepsilon_1\Lambda \cong R$, $\varepsilon_{26}\Lambda \cong R^{26\times 26}$, $\varepsilon_{26'}\Lambda \cong R[\sqrt{-2}]^{26\times 26}$ *and*
$\varepsilon_{13}\Lambda \cong \Lambda(1,12,\begin{pmatrix}0&2\\0&0\end{pmatrix})$, $\varepsilon_{27}\Lambda \cong \Lambda(1,26,\begin{pmatrix}0&2\\0&0\end{pmatrix})$ *and*
$\varepsilon_{39}\Lambda \cong \Lambda(1,12,26,\begin{pmatrix}0&2&2\\0&0&2\\0&2&0\end{pmatrix})$.

Proof: G has 5 blocks, 4 of which are of defect zero, and 7 2'-conjugacy classes. Hence there remain 3 Brauer characters for the principal block, one of which is 1 . (The character table of G is in [McK 79].) Restricting the Frobenius character 12 to the 13-Sylow subgroup yields the character of an irreducible (though not absolutely irreducible) module, since 2 is a primitive root mod 13. Since all Frobenius characters in Λ take rational values on the classes with element order 13 (not depending on the class) this implies that the restriction of 12 to the 2'-classes yields an absolutely irreducible Brauer character. A similar argument applied to the Frobenius character 26 with the Sylow-3-subgroup yields that the third irreducible Brauer character is the restriction of 26 to 2'-classes. Therefore the decomposition matrix of the principal block of R'G is as given above, where R' is a sufficiently big unramified extension of R . However, the irreducible Brauer characters take rational values on all 2'-classes. Hence $F = R/\mathfrak{p}$ is a splitting field of $\Lambda/\mathfrak{p}\Lambda$. Moreover, the values of the central character of 26' generate the maximal order

$R[\sqrt{-2}]$ of $K[\sqrt{-2}]$. Hence by (III.13) the Schur indices of all $\varepsilon_s\Lambda$ are equal to 1 , and then, by (II.8) and the comments following (II.8), one concludes that all $\varepsilon_s\Lambda$, $s = 1, 12, 13, 26, 26', 27, 36$, are graduated orders and that the decomposition matrix of Λ is the one given above. Let $\begin{pmatrix} 0 & a \\ 0 & 0 \end{pmatrix}$ be the exponent matrix of $\varepsilon_{13}\Lambda$, $\begin{pmatrix} 0 & b \\ 0 & 0 \end{pmatrix}$ the one of $\varepsilon_{27}\Lambda$, and $\begin{pmatrix} 0 & x & y \\ 0 & 0 & z \\ 0 & z' & 0 \end{pmatrix}$ the one of $\varepsilon_{39}\Lambda$. Then the amalgamation matrices of the projective indecomposable Λ-lattices P_1 and P_{12} are:

$\alpha(P_1)$	1	12	26
1	4	0	0
13	4	4-a	0
27	4	0	4-b
39	4	4-x	4-y

$\alpha(P_{12})$	1	12	26
12	0	2	0
13	4-a	4	0
39	4-x	4	4-z-z'

Hence (by (IV.7)) $4-a = 4-x$, $4-b = 4-y$, and $4-z-z' = 0$. Therefore $x = a$ $y = b$ and $z+z' = 4$.

The first claim is $a = 2$. Since the central characters of G belonging to 1 and 13 are congruent modulo 4 but not modulo 8 , (IV.13) implies $a \leq 2$. The Frobenius character 13 is induced up from a one-dimensional rational character of a subgroup. Hence one of the lattices L belonging to 13 is monomial, i.e. has a basis $l_1,\ldots,l_{13}$ which is permuted with sign changes by the operation of G . $L_1 = \{\sum_{i=1}^{13} a_i l_i \mid a_i \in R , \sum_{i=1}^{13} a_i \in \mathfrak{p}\}$ is an RG-sublattice and $L_1^{\#} = \{\frac{1}{2}\sum_{i=1}^{13} a_i l_i \mid a_i \in R , a_i \equiv a_j \pmod{\mathfrak{p}} \quad 1 \leq i,j \leq 13\}$ an RG-lattice containing L . Since $L_1^{\#}/L_1 \underset{RG}{\cong} R/\mathfrak{p}^2$, one gets $a \geq 2$. Hence $a = 2$.

On the other hand the Frobenius character 39 and all three Brauer characters are real. Hence by (IV.1) one gets $z - z' = y - x$. Together with $x = a = 2$ and the above equation for $z + z'$, one gets $z = \frac{1}{2}(y+2)$, $z' = \frac{1}{2}(6-y)$. Hence y is even, and therefore $y = 2$ or

$y = 4$.

The central characters belonging to 1 and 27 are congruent modulo 8 , but not modulo 16 . Hence by (IV.13) $b \leq 3$. Since $b = y$, one now gets $y = 2$. Altogether, one has $x = a = y = b = z = z' = 2$, as claimed.

q.e.d.

In the next example the intersection of two sets c_i has cardinality bigger than 2 in one instance. The difficulties are easily overcome because of an outer automorphism of order 3 . The defect is 2 .

(VI.5) Example. Let $G = PSL_3(4)$, R the ring of 3-adic integers, and Λ the principal block ideal of RG . The decomposition numbers are given by

	1	19	15_1	15_2	15_3
1	1				
20	1	1			
35_1	1	1	1		
35_2	1	1		1	
35_3	1	1			1
64		1	1	1	1

One has $\varepsilon_1\Lambda \cong R$, $\varepsilon_{20}\Lambda = \Lambda(1,19,\begin{pmatrix}0&1\\0&0\end{pmatrix})$, $\varepsilon_{35_i}\Lambda \cong \Lambda(1,19,15,\begin{pmatrix}0&1&2\\0&0&1\\0&0&0\end{pmatrix})$,

and $\varepsilon_{64}\Lambda \cong \Lambda(19,15,15,15,\begin{vmatrix}0&1&1&1\\0&0&1&1\\0&1&0&1\\0&1&1&0\end{vmatrix})$.

Proof: From the character table, cf. e.g. [McK 79] one sees that 5 irreducible Brauer characters of G belong to Λ . The decomposition matrix is obtained by inducing up the projective indecomposable $RGL_3(2)$-lattices to G , (note $GL_3(2) \leq G$), taking the direct summands in Λ , and using the symmetry induced by an outer automorphism of order three (note $G \leq PGL_3(4)$). Again the Brauer characters take only rational values, the Schur indices are equal to 1 , and all $\varepsilon_s\Lambda$ are graduated orders (in $K^{l_s \times l_s}$). G has a 2-transitive permutation representation of degree 21 (= $(4^3-1)\cdot(4-1)^{-1}$). Therefore by the decomposition matrix and (VI.1) one obtains $\begin{pmatrix} 0 & 1 \\ 0 & 0 \end{pmatrix}$ as the exponent matrix of $\varepsilon_{20}\Lambda$. Let $\begin{pmatrix} 0 & a & b \\ 0 & 0 & c \\ 0 & c' & 0 \end{pmatrix}$ be the exponent matrix of $\varepsilon_{35_1}\Lambda$; (the Brauer characters are arranged as in the decomposition matrix.) Because of the outer automorphism of order 3 of G this is also the exponent matrix for the other two $\varepsilon_{35_i}\Lambda$. Finally let $\begin{pmatrix} 0 & x & y & z \\ 0 & 0 & u & v \\ 0 & u' & 0 & w \\ 0 & v' & w' & 0 \end{pmatrix}$ be the exponent matrix of $\varepsilon_{64}\Lambda$. Since $c_{15_i} \cap c_{15_j} = \{64\}$ for $1 \leq i < j \leq 3$, one gets from (IV.7) $u+u' = v+v' = w+w' = 2$.

The defect is two and all characters are real. Hence, the remarks in the proof of (VI.3) on the relation between the first line of an exponent matrix and the rest of the same exponent matrix can be applied. Hence $u = u' = v = \ldots = w' = 1$ and $x = y = z$. Since $c_1 \cap c_{15_1} = \{35_1\}$, (IV.7) implies $b = 2$. The amalgamation matrix of P_1 is

$\alpha(P_1)$	1	19	15_1	15_2	15_3
1	2				
20	2	1			
35_1	2	2-a	2-b		
35_2	2	2-a		2-b	
35_3	2	2-a			2-b

Hence (for example by (IV.7) (iii)) $1 = 2-a$, i.e. $a = 1$, hence $c = 1$ $c' = 0$. Since $c_{19} \cap c_{15_1} = \{35_1, 64\}$ one now gets $x = 1$ ($= y = z$).

q.e.d.

It is possible that the principal 3-blocks of $PSL_3(q)$ for $q \equiv 4$ mod 9 has the same decomposition matrix and that the same argument as above leads to the same exponent matrices as above. If this is so one might suspect that the blocks (over some unramified extension of the 3-adic integers) are Morita equivalent. The last examples are concerned with the Mathieu group M_{11} at the two (non-trivial) primes 2 and 3. Again only the principal blocks are of interest. The defects are 4 and 2 respectively.

(VI.6) Example. Let $G = M_{11}$, R the ring of 2-adic integers, and Λ the principal block ideal of RG. The decomposition numbers in the modified form as defined in (III.10) are

	1	10	44
1	1		
10		1	
10'		1	
11	1	1	
44			1
45	1		1
55	1	1	1

where the center of $\varepsilon_{10'}A$ is isomorphic to $K[\sqrt{-2}]$. One has $\varepsilon_1\Lambda \cong R$, $\varepsilon_{10}\Lambda \cong R^{10\times 10}$, $\varepsilon_{10'}\Lambda \cong R\sqrt{-2}^{\,10\times 10}$, $\varepsilon_{11}\Lambda \cong \Lambda(1,10,\begin{pmatrix} 0 & 2 \\ 0 & 0 \end{pmatrix})$, $\varepsilon_{45}\Lambda \cong \Lambda(1,44,\begin{pmatrix} 0 & 2 \\ 0 & 0 \end{pmatrix})$, and $\varepsilon_{55}\Lambda \cong \Lambda(1,10,44,\begin{pmatrix} 0 & 2 & 2 \\ 0 & 0 & 2 \\ 0 & 2 & 0 \end{pmatrix})$.

(Note, the graduated hull is Morita equivalent to the graduated hull in (VI.4). I have not checked whether the rings itself are Morita equivalent.)

Proof: The usual decomposition matrix can be found in [Jam 73], the same arguments as in (VI.4) show that the (modified) decomposition numbers are as above, that all Schur indices are equal to 1 and that the $\varepsilon_s\Lambda$ are graduated orders. That the exponent of $\varepsilon_{11}\Lambda$ is $\begin{pmatrix} 0 & 2 \\ 0 & 0 \end{pmatrix}$ follows from (VI.1) via the permutation representation of degree 12 . The congruences of the central character belonging to 1 and 45 show that the exponent matrix of $\varepsilon_{45}\Lambda$ is $\begin{pmatrix} 0 & a \\ 0 & 0 \end{pmatrix}$ with $a \leq 3$. The rest of the argument is analogous to the one for (VI.4).

q.e.d.

The last example is the most complicated one in this chapter, and uses the results of an electronic computation of the submodules of two FG-modules.

<u>(VI.7) Example:</u> Let $G = M_{11}$, R the ring of 3-adic integers, and Λ the principal block ideal of RG . The decomposition numbers of Λ are given by

	1	10	$\overline{10}$	5	$\overline{5}$	$10'$	24
1	1						
10		1					
$\overline{10}$			1				
16	1	1		1			
$\overline{16}$	1		1		1		
$10'$						1	
11	1			1	1		
44				1	1	1	1
55	1	1	1	1	1		1

One has $\varepsilon_1\Lambda \cong R$, $\varepsilon_{10}\Lambda \cong \varepsilon_{\overline{10}}\Lambda \cong \varepsilon_{10'}\Lambda \cong R^{10\times 10}$,

$$\varepsilon_{16}\Lambda \cong \Lambda(1, 10, 5, \begin{pmatrix} 0 & 1 & 1 \\ 0 & 0 & 1 \\ 0 & 0 & 0 \end{pmatrix}) , \quad \varepsilon_{\overline{16}}\Lambda \cong \Lambda(1, \overline{10}, \overline{5}, \begin{pmatrix} 0 & 1 & 1 \\ 0 & 0 & 0 \\ 0 & 1 & 0 \end{pmatrix}) .$$

(The symbols $\overline{10}$, $\overline{5}$ in the standard notation for graduated orders indicate which components of $\Lambda/Jac(\Lambda)$ are mapped onto those of $\varepsilon_s\Lambda/Jac(\varepsilon_s\Lambda)$).

$$\varepsilon_{11}\Lambda \cong \Lambda(1,5,\overline{5},\begin{pmatrix}0&1&1\\0&0&0\\0&1&0\end{pmatrix}) , \quad \varepsilon_{44}\Lambda \cong \Lambda(5,\overline{5},10',24,\begin{pmatrix}0&1&2&1\\0&0&2&0\\0&0&0&0\\0&1&2&0\end{pmatrix}) ,$$

$$\text{and} \quad \varepsilon_{55}\Lambda \cong \Lambda(1,10,\overline{10},5,\overline{5},24,\begin{pmatrix}0&1&1&1&1&2\\0&0&1&1&1&2\\0&1&0&1&0&1\\0&0&1&0&0&1\\0&1&1&1&0&1\\0&0&1&0&0&0\end{pmatrix}) .$$

Proof: All character values lie in K. The decomposition numbers are determined in [Jam 73]. As usual one concludes that the modified decomposition numbers are as above, that all Schur indices are equal to 1 and that the $\varepsilon_s\Lambda$ are graduated orders in $K^{1_s \times 1_s}$.

Let $\begin{pmatrix}0&a_1&a_2\\0&0&b_1\\0&b_2&0\end{pmatrix}$ be the exponent matrix of $\varepsilon_{11}\Lambda$.

Since S_5 and $S_{\overline{5}}$ are dual to each other, one obtains from (IV.1) $a_1 = a_2$. Since 11 is constituent of the 2-transitive permutation representation of M_{11} on 12 points, (VII.1) implies $a_1 = a_2 = 1$. Hence $b_i \leq 1$ for $i = 1,2$ by (II.2) (*). By restricting the operation of G to the subgroup isomorphic to $PSL_2(11)$, one gets $b_1 + b_2 = 1$ (note, $9 \nmid |PSL_2(11)|$). It is now a matter of choosing the notation to decide $b_1 = 0$ and $b_2 = 1$.

Let $L = \varepsilon_{11}P_1$, $L' \leq L$ with $L/L' \cong S_1$, and $L'' \leq L'$ with $L'/L'' \cong S_{\overline{5}}$. Two facts about the 3-modular representation theory of G will be used, namely the lattices of submodules of the exterior squares $\Lambda^2(L/\mathfrak{p}L)$ and $\Lambda^2(L'/\mathfrak{p}L')$. They can conveniently be computed on a computer by the Parker-algorithm for reducing modular representations, cf. [Par 82]; the computations have been carried out by D. Ruland on the Cyber 175 in Aachen. To formulate the results of these computations note that $N = \Lambda^2(L)$ is an irreducible Λ-lattice of dimension 55, as easily seen from the character table. Identify $R^{55\times 55}$ with $\mathrm{End}_R(N)$. Then (in obvious notation using the embedding of $\varepsilon_{55}\Lambda$ in $\mathrm{End}_R(N)$ cf. comments after (II.6)):

(i) $\varepsilon_{55}\Lambda + \mathrm{End}_R(N) \cong \Lambda(1,10,\overline{10},5,\overline{5},24, \begin{pmatrix} 0&1&0&0&0&1\\ 0&0&0&0&0&1\\ 1&1&0&1&0&1\\ 1&1&1&0&0&1\\ 1&1&1&1&0&1\\ 1&1&1&0&0&0 \end{pmatrix})$;

and, identifying $\mathrm{End}_R(N')$ with $R^{55\times 55}$, where $N' = \Lambda^2(L')$, then (in the notation above via the embedding of $\varepsilon_{55}\Lambda$ into $\mathrm{End}_R(N')$,

(ii) $\varepsilon_{55}\Lambda + \mathrm{End}_R(N') = \Lambda(1,10,\overline{10},5,\overline{5},24, \begin{pmatrix} 0&1&0&1&1&1\\ 0&0&0&1&1&1\\ 1&1&0&1&1&1\\ 0&0&0&0&0&0\\ 0&1&0&1&0&0\\ 1&1&1&1&1&0 \end{pmatrix})$.

Cf. also fig. 3

By taking the intersection of these two orders after embedding the second one into $\mathrm{End}_R(N)$, which is identified with $R^{55\times 55}$ (note, $N' = L((0,0,0,1,1,0))$.) one obtains,

$$\varepsilon_{55}\Lambda \subseteq \Lambda(1,10,\overline{10},5,\overline{5},24,X) \text{ with } X = \begin{pmatrix} 0&1&0&0&0&1\\ 0&0&0&0&0&1\\ 1&1&0&1&0&1\\ 1&1&1&0&0&1\\ 1&2&1&1&0&1\\ 1&1&1&0&0&0 \end{pmatrix} = (x_{ij})_{i,j\in r_{55}}$$

If $M = (m_{ij})_{i,j\in r_{55}}$ denotes the exponent matrix of $\varepsilon_{55}\Lambda$ with respect to the embedding into $\mathrm{End}_R(N)$, then this inclusion means $m_{ij} \geq x_{ij}$ for $i,j \in r_{55}$.

Moreover, (what is an even more valuable information), equality holds for those positions $(i,j) \in r_{55}\times r_{55}$ where one has a zero in the exponent matrix under (i) or under (ii), (cf. remarks before (II.22)). Since the defect of the block is 2, one also has $m_{ij} = x_{ij}$ and $m_{ji} = x_{ji}$ whenever $x_{ij} + x_{ji} = 2$.

Since $c_{10} \cap c_{\overline{10}} = \{55\}$, (IV.7) implies $m_{10,\overline{10}} + m_{\overline{10},10} = 2$; and since $m_{10,\overline{10}} = 0$, one has $m_{\overline{10},10} = 2$ ($\neq x_{10,\overline{10}}$).

Now, $m_{1,10}$, $m_{\overline{10},1}$, and $m_{\overline{5},5}$ are the only entries of M which are not yet determined. From $m_{1,10} = 2$ one obtains a contradiction by (II.2) (*) to $m_{1,5} = 0$ and $m_{10,5} = 1$. Hence $m_{1,10} = 1$.

(IV.1) implies $m_{1,10,1} = m_{1,\overline{10},1}$, and hence $m_{\overline{10},1} = m_{1,10} = 1$. Finally $m_{\overline{5},5} = 1$ can be proved as follows: Elementary manipulations with exterior squares show that N/N' and $L'/\mathfrak{p}L$ are isomorphic. Hence by (IV.11) S_5 is a composition factor of the amalgamating factor of $(\varepsilon_{11} + \varepsilon_{55})P_{\overline{5}}$ as amalgam of $\varepsilon_{11}P_{\overline{5}} \cong L'$ and $\varepsilon_{55}P_{\overline{5}} \cong N$. Hence $2 - m_{5,\overline{5}} - m_{\overline{5},5} = 1$, i.e. $m_{\overline{5},5} = 1$. Transforming M in such a way that the first column becomes zero, cf. (II.7) ff., yields the exponent matrix as claimed above. The lattice of submodules of $N = \Lambda^2(L)$ is given in the following diagram: $(N' = \Lambda^2(L')$, $N'' = \Lambda^2(L''))$:

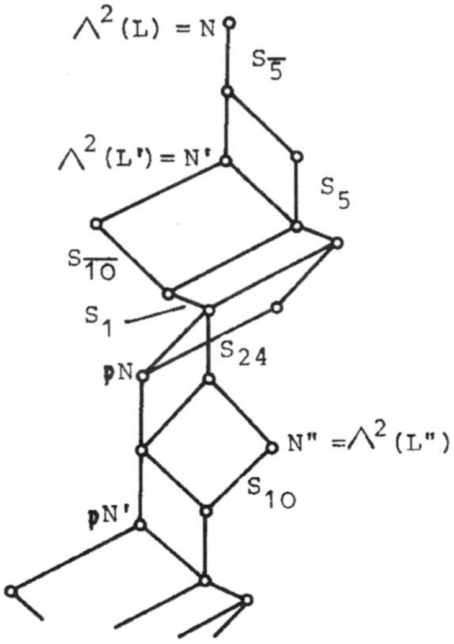

fig. 3

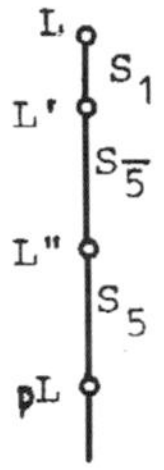

fig. 4

Now, let $\begin{pmatrix} 0 & a & b \\ 0 & 0 & c \\ 0 & d & 0 \end{pmatrix}$ be the exponent matrix of $\varepsilon_{16}\Lambda$. By (IV.1) the exponent matrix of $\varepsilon_{\overline{16}}\Lambda$ is equal to $\begin{pmatrix} 0 & a & b \\ 0 & 0 & d-a+b \\ 0 & c-b+a & 0 \end{pmatrix}$. For the exponent matrix for $\varepsilon_{44}\Lambda$ it is convenient to let the third column (corresponding to 10') consist of zeroes (instead of the first one). Then the non-diagonal entries of the third row are all equal to 2 by

(IV.7). Because of the standard anti-isomorphism of Λ the exponent matrix of $\varepsilon_{44}\Lambda$ then is of the form

$$\begin{pmatrix} 0 & u & 0 & x \\ v & 0 & 0 & y \\ 2 & 2 & 0 & 2 \\ y & x & 0 & 0 \end{pmatrix} \quad \text{by (IV.1).}$$

The amalgamation matrices of P_1 , $P_{\bar{5}}$, and P_{24} are

$\alpha(P_1)$	1	10	$\overline{10}$	5	$\bar{5}$	10'	24
1	2						
16	2	2-a		2-b			
$\overline{16}$	2		2-a		2-b		
11	2			1	1		
55	2	1	1	1	1		0

$\alpha(P_{\bar{5}})$	1	10	$\overline{10}$	5	$\bar{5}$	10'	24
$\overline{16}$	2-b		2-c-d		2		
11	1			1	2		
44				2-u-v	2	0	2-x-y
55	1	0	1	1	2		1

$\alpha(P_{24})$	1	10	$\overline{10}$	5	$\bar{5}$	10'	24
44				2-x-y	2-x-y	0	2
55	0	0	0	1	1		2

(The empty positions have to be filled up with zeroes.) From $\alpha(P_1)$ and $\alpha(P_{\bar{5}})$ one gets $a = 1$, $c + d = 1$.
A closer analysis of $P_{\bar{5}}$ shows $b = 1$: Assume $b = 2$. Then S_1 is composition factor of the amalgamating factor of $(\varepsilon_{11} + \varepsilon_{55})P_{\bar{5}}$ as amalgam of $\varepsilon_{11}P_{\bar{5}} \cong L'$ and $\varepsilon_{55}P_{\bar{5}} \cong N$. But this is impossible as a comparison of the lattices of Λ-sublattices of L' and N shows, cf. figs. 3 and 4. Hence $b = 1$. Moreover, one gets $c = 1$ and $d = 0$ from an analysis of $\alpha(P_1)$ (or $\alpha(P_{\bar{5}})$). Since there are only two entries $\neq 0$ in the column 10 of $\alpha(P_1)$, one has S_{10} as composition factor of the

amalgamating factor of $(\varepsilon_{16}+\varepsilon_{55})P_1$. From the lattice of Λ-sublattices of $\varepsilon_{55}P_1$ (cf. fig. 3) one concludes that $Jac(\varepsilon_{16}P_1)$ has a maximal sublattice X with $Jac(\varepsilon_{16}P_1)/X \cong S_{10}$. Hence $c=1$ and $d=0$ (note, $\varepsilon_{16}P_1$ was already known to be uniserial by $a=b=c+d=1$.).

Finally the exponent matrix of $\varepsilon_{44}\Lambda$ will be determined.

From $\alpha(P_{\bar{5}})$ or $\alpha(P_{24})$ one gets $x+y=1$.

Denote the irreducible $\varepsilon_{44}\Lambda$-lattice $\varepsilon_{44}P_{10'}$ by X . Then $\varepsilon_{44}P_{24}$ is isomorphic to $X((x,y,2,0)^{tr})$. From the amalgamation matrix $\alpha(P_{24})$ and the lattice of submodules of $\varepsilon_{55}P_{24}$ (cf. fig. 3), one sees that $X((x,y,2,0)^{tr})$ has a sublattice $X((x+1,y+1,2,2)^{tr})$ such that the factor module is uniserial with composition factors S_{24}, $S_{\bar{5}}$, S_5, S_{24} (in the order from top to bottom.). In particular, X has no sublattice of the form $X((x+1,y,2,1)^{tr})$. Using (**) in (II.4) and the existence of $X((x,y,2,1)^{tr})$ one gets $u+y=x$. Since $x+y=1$, one concludes that u is odd, and hence $u=1$, $y=0$, $x=1$. Now (*) in (II.2) yields $v=0$. Transforming the exponent matrix in such a way that the first column is zero, cf. (II.6), yield the claimed exponent matrix.

q.e.d.

VII. The principal 2-block of $SL_2(q)$ for odd prime powers q

In this chapter a description of the isomorphism type of the principal block ideal Λ of RG will be given, where $G = SL_2(q)$, q an odd prime power and where $R(=\mathbb{Z}_2)$ is the ring of 2-adic integers. This problem is divided into two smaller problems: Let $\bar{G} = PSL_2(q)$ and $\langle z\rangle$ be the center of $G = SL_2(q)$. Then $E = \frac{1}{2}(1-z) \in KG$ ($K = \mathbb{Q}_2$; the notation of Chapter IV is kept) and Λ has two epimorphic images $\Lambda_1 := (1-E)\Lambda$ and $\Lambda_2 := E\Lambda$, which are selfdual orders; namely Λ_1 is the principal block ideal of $R\bar{G}$ and by (III.4) or (III.2) Λ_2 can be viewed as a block ideal of a twisted group ring of $\bar{G}$. Moreover Λ can be recovered from Λ_1 and Λ_2 as a pullback of the (obvious) epimorphisms of Λ_1 and Λ_2 onto the principal block ideal b ($\cong \Lambda_1/\mathfrak{p}\Lambda_1 \cong \Lambda_2/\mathfrak{p}\Lambda_2$) of $F\bar{G}$, i.e. Λ is a subdirect product of Λ_1 and Λ_2 with b as amalgamating factor. It will turn out that the structure of Λ_2 is easy to obtain and to describe. Therefore one has a description of b which can be used for the investigation of Λ_1 . As a corollary one obtains the description of the submodule structure of the projective indecomposable b-modules given by Erdmann, cf. [Erd 77] and also by Alperin (unpublished, cf. also [Alp 72]). It should be noted that the decomposition numbers are not always equal to 0 or 1 in case $q \equiv 1 \bmod 4$. Moreover, the cases $q \equiv 3$ and $5 \bmod 8$ need special treatments, since in these cases the endomorphism fields of some simple Λ-modules are bigger than F .

The following notation will be used:

If S_1 and S_2 are two orders with ideals $\mathfrak{a}_i \trianglelefteq S_i$ $(i=1,2)$ such that there are epimorphisms $\nu_i : S_i \to X$ onto some ring X with kernels $\mathfrak{a}_i$, then

$$S_1^{n\times n} \overset{(\mathfrak{a}_1,\mathfrak{a}_2)}{\rule{3cm}{0.4pt}} S_2^{n\times n} = S_1^{n\times n} \underset{(\mathfrak{a}_1,\mathfrak{a}_2)}{\rule{3cm}{0.4pt}} S_2^{n\times n}$$

denotes the matrix ring of degree n over

$\{(x_1,x_2) \in S_1 \oplus S_2 \mid \nu_1(x_1) = \nu_2(x_2)\}$. Sometimes, in particular if $S_1 = S_2$ and $\mathfrak{a}_1 = \mathfrak{a}_2$, the symbol $(\mathfrak{a}_1,\mathfrak{a}_2)$ will be replaced by $\mathfrak{a}_1$ or $\mathfrak{a}_2$, as done earlier in this paper. Note, the above type of ring will usually be a subring of a bigger ring and therefore the symbol will be only part of the full description of the ring under discussion; also the line connecting the S_i might not be straight. Further notation:

$R, K = \mathbb{Q}_2$, F, $\mathfrak{p}$, G, $\bar{G}$, E, Λ, Λ_1, Λ_2, b, A as just defined (cf. also Chapter IV), $A^1 = (1-E)A = K\Lambda_1$, $A^2 = EA = K\Lambda_2$; $\tau := \frac{q-1}{2}$.
The ($\mathbb{C}$-)irreducible Brauer characters of Λ will be denoted by their degrees: 1, τ_1, τ_2 , cf. [Bra 66].
$K' (:= K[x]/(x^2+x+1))$ denotes the unique unramified extension of degree 2 of K, R' the maximal order of K', $\mathfrak{p}' = \mathfrak{p}R'$,
$F' = R'/\mathfrak{p}'$, $\Lambda' = R' \otimes_R \Lambda$, $\Lambda_1' = R' \otimes_R \Lambda_1$, ... , $A_2' = K' \otimes_K A_2$, etc.
The absolutely irreducible Frobenius characters in Λ are denoted by their degrees d . The same symbol d is used for the irreducible A-module of this (set of) character(s), and the central corresponding primitive idempotent of A is denoted by ε_d .

From the character table of G , cf. [Sch 07], or [Dor 72], one sees that the only possible irrationalities of the two Brauer characters τ_1, τ_2 are $\frac{-1 \pm \sqrt{\varepsilon q}}{2}$ with $\varepsilon = (-1)^\tau$. Since they are roots of $x^2 + x + \frac{1-\varepsilon q}{4}$, the minimal modular splitting field for $\Lambda/2\Lambda$ is $F(\cong \mathbb{Z}/2\mathbb{Z})$ in case $q \equiv \pm 1$ (8) and $F'(\cong F_4)$ in case $q \equiv \pm 3$ (8) .
These four cases will be treated separately.

<u>Case i:</u> $q \equiv 3 \bmod 8$
$\tilde{\tilde{R}} \otimes_R \Lambda_i$ for $\tilde{\tilde{R}}$ sufficiently big was investigated in (V.5) for $i = 1$ and in (V.6) for $i = 2$.

<u>*(VII(.1) Theorem.*</u> *Let $q \equiv 3 \bmod 8$, $\tau = \frac{q-1}{2}$, and R the ring of 2-adic integers.*

(i) The principal block ideal Λ_1 *of* $R\bar{G}$, $\bar{G} = PSL_2(q)$, *is iso-morphic to the R-order*

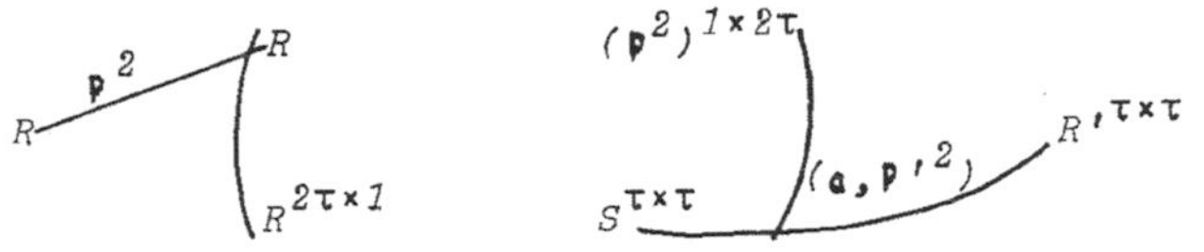

$(\subseteq \Lambda(1,(0)) \oplus \Lambda(1,2\tau, \begin{pmatrix} 0 & 2 \\ 0 & 0 \end{pmatrix}) \oplus \Lambda(R',\tau,(0)))$

with $S = \mathfrak{p}^{2\times 2} + R[\begin{pmatrix} 0 & 1 \\ 1 & 1 \end{pmatrix}] \subseteq R^{2\times 2}$, *(note,* $R[\begin{pmatrix} 0 & 1 \\ 1 & 1 \end{pmatrix}] \cong R')$,

and $\mathfrak{a} = \mathfrak{p} \cdot \langle \begin{pmatrix} 1 & 1 \\ 0 & 1 \end{pmatrix}, \begin{pmatrix} 1 & 0 \\ 1 & 1 \end{pmatrix}, \mathfrak{p}^{2\times 2} \rangle_R = \mathfrak{p}^2 R[\begin{pmatrix} 0 & 1 \\ 1 & 1 \end{pmatrix}] + \mathfrak{p}\begin{pmatrix} 0 & 1 \\ 1 & 0 \end{pmatrix} R[\begin{pmatrix} 0 & 1 \\ 1 & 1 \end{pmatrix}] \lhd S.$

(ii) The epimorphic image Λ_2 *of the principal block* Λ *of* RG , $G = SL_2(q)$, *is isomorphic to the R-order*

$$\begin{pmatrix} \tilde{R} & \mathfrak{p}'^{1\times\tau} \\ R'^{\tau\times 1} & R'^{\tau\times\tau} \end{pmatrix} \underline{\quad(\mathfrak{p}',\mathfrak{P})\quad} \Omega^{\tau\times\tau} \quad (\subseteq \Lambda(R',1,\tau,\begin{pmatrix} 0 & 1 \\ 0 & 0 \end{pmatrix}) \oplus \Lambda(\Omega,\tau,(0)))$$

with $\tilde{R} = \mathfrak{p}' + R1 \subseteq R'$, *and where* Ω *is the maximal order in the unique central division algebra* D *over* K *with index* 2, $\mathfrak{P} = Jac(\Omega)$.

Proof: (i). Either directly or by applying (III.13) one gets $A^1 \cong K \oplus K^{q\times q} \oplus K'^{\tau\times\tau}$. From the decomposition numbers, cf. (V.5) and the above remarks on the splitting fields of the simple Λ_1-modules one has $\varepsilon_1\Lambda_1 \cong R$ and $\varepsilon_\tau\Lambda_1 \cong R'^{\tau\times\tau}$ by (III.12) or (II.5). Moreover by

(V.5) $R' \otimes_R \varepsilon_q\Lambda_1 \cong \Lambda(R',1,\tau,\tau, \begin{pmatrix} 0 & 2 & 2 \\ 0 & 0 & 1 \\ 0 & 1 & 0 \end{pmatrix})$. Since

$\varepsilon_q\Lambda_1/Jac(\varepsilon_q\Lambda_1) \cong F \oplus F'^{\tau\times\tau}$, the graduated hull of $\varepsilon_q\Lambda_1$ is isomorphic to $\Lambda(1,2\tau, \begin{pmatrix} 0 & 2 \\ 0 & 0 \end{pmatrix})$. Now $\varepsilon_q\Lambda_1 \cong \begin{pmatrix} R & (\mathfrak{p}^2)^{1\times 2\tau} \\ R^{2\tau\times 1} & S^{\tau\times\tau} \end{pmatrix} =: \Gamma$ follows easily,

e.g. from (II.20). The rest is straightforward by using (III.8); for

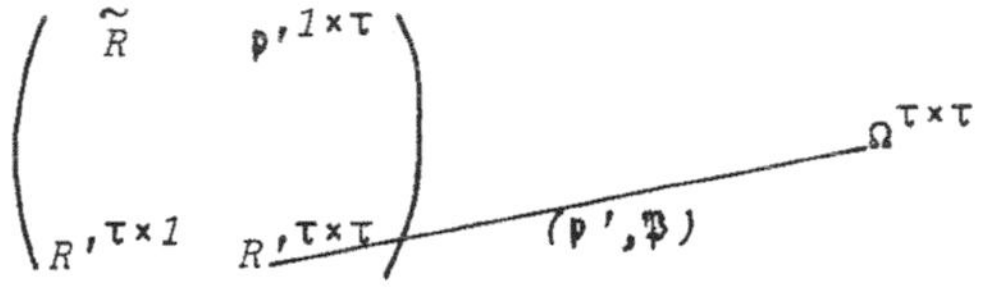

the computation of a the selfduality of $S \underset{(a,\mathfrak{p}'^2)}{\text{———}} R'$, cf. (III.6), or (II.17) is used.

(ii). (III.13) and (V.6) imply $A^2 \cong K'^{(\tau+1)\times(\tau+1)} \oplus D^{\tau\times\tau}$ (the centers K' and K of these components can be obtained from the character table of G), $\varepsilon_{q-1}\Lambda_2 \cong \Omega^{\tau\times\tau}$ and the appearance of $R'^{\tau\times\tau}$ in $\varepsilon_{\tau+1}\Lambda_2$. By using $\varepsilon_{\tau+1}\Lambda_2$ / Jac $(\varepsilon_{\tau+1}\Lambda_2) \cong F \oplus F'^{\tau\times\tau}$ (cf. (V.6) and part (i) of this proof), one obtains the desired description of $\varepsilon_{\tau+1}\Lambda_2$ and also of Λ_2 by the same argument as in (i).

q.e.d.

(VII.2) Corollary (cf. [Erd 77] for F algebraically closed)
Let X_1 and X_τ be the b-modules which are the projective covers of the two simple b-modules S_1 and S_τ where $b \cong \Lambda_1/\mathfrak{p}\Lambda_1 \cong \Lambda_2/\mathfrak{p}\Lambda_2$ is the principal (2-)block of $F\bar{G}$. Then X_1 is uniserial with composition factors S_1, S_τ, S_1, and X_τ satisfies $Jac(X_\tau)/Soc(X_\tau) \cong S_1 \oplus S_1$.

Case ii: $q \equiv 5$ mod 8.

(VII.3) Lemma. For $q \equiv 5 \pmod 8$ K' is a splitting field for A^1 and the decomposition numbers for Λ' are given by

	1	τ_1	τ_2
1	1	0	0
q	1	1	1
$(1+\tau)_1$	1	1	0
$(1+\tau)_2$	1	0	1
τ_1	0	1	0
τ_2	0	0	1
$q+1$	2	1	1

where the part of the matrix above the dotted line refers to Λ_1' and the one below to Λ_2' .

(i) The exponent matrices for $\varepsilon_s\Lambda_1'$ *are given by* (0),

$$\begin{pmatrix} 0&1&1\\ 0&0&1\\ 0&1&0 \end{pmatrix}, \quad \begin{pmatrix} 0&1\\ 0&0 \end{pmatrix}, \text{ and } \begin{pmatrix} 0&1\\ 0&0 \end{pmatrix}$$ *(with the Brauer- and Frobenius character ordered as above.)*

(ii) $\varepsilon_{\tau_1}\Lambda_2' \cong \varepsilon_{\tau_2}\Lambda_2' \cong R'^{\tau\times\tau}$ *and* $\varepsilon_{q+1}\Lambda_2' \cong$

$$\begin{pmatrix} R' & \mathfrak{p}'^{1\times\tau} & \mathfrak{p}' & \mathfrak{p}'^{1\times\tau}\\ R'^{\tau\times 1} & R'^{\tau\times\tau} & \mathfrak{p}'^{\tau\times 1} & \mathfrak{p}'^{\tau\times\tau}\\ R' & R'^{1\times\tau} & R' & \mathfrak{p}'^{1\times\tau}\\ R'^{\tau\times 1} & R'^{\tau\times\tau} & R'^{\tau\times 1} & R'^{\tau\times\tau} \end{pmatrix} \quad (\subseteq \Lambda(R', 1, \tau, 1, \tau, H_4) \text{ with } H_4 = \begin{pmatrix} 0&1&1&1\\ 0&0&1&1\\ 0&0&0&1\\ 0&0&0&0 \end{pmatrix})$$

$\mathfrak{p}'$

Proof: The decomposition numbers of $\tilde{R}\otimes_R\Lambda$ for $\tilde{R}$ sufficiently big follow from the Brauer characters and the character table of G. That K' is a splitting field (and that therefore $\tilde{R}$ can be chosen to be R'), follows as usual from (III.13), since F' is a splitting field for $\Lambda/\mathfrak{p}\Lambda$.

(i) Since $\bar{G}$ has an outer automorphism interchanging the two Brauer characters τ_1 and τ_2 (induced by the action of $PGL_2(q)$) the exponent matrices of the $\varepsilon_s\Lambda_1'$ are of the form

$$(0), \quad \begin{pmatrix} 0&a&a\\ 0&0&b\\ 0&b&0 \end{pmatrix}, \quad \begin{pmatrix} 0&c\\ 0&0 \end{pmatrix} \text{ and } \begin{pmatrix} 0&c\\ 0&0 \end{pmatrix} \quad \text{for suitable } a,b,c \in \mathbb{Z}_{>0}$$

(cf. IV.1). The amalgamation matrices for projective Λ_1'-lattices yield $a = c$ and $b = 1$ (cf. (IV.7); note, the Sylow 2-subgroups of $\bar{G}$ have order 2^2). Finally $a = 1$ follows from (VI.1) via the embedding of $\bar{G}$ into the symmetric group of degree $q+1$ (note, $4\nmid q+1$).

(ii) The argument for the investigation of Λ_2' is exactly the same as for Λ_2 in the case $q \equiv 1 \bmod 8$ in (VII.10) and is therefore omitted.

q.e.d.

<u>(VII.4) Theorem.</u> *Let* $q \equiv 5 \bmod 8$, $\tau = \frac{q-1}{2}$, *and* R *the ring of 2-adic integers.*

(i) The principal block ideal Λ_1 *of* $R\bar{G}$, $\bar{G} = PSL_2(q)$, *is isomorphic to the R-order with unique graduated hull* $R \oplus \Lambda(1, 2\tau, \begin{pmatrix} 0 & 1 \\ 0 & 0 \end{pmatrix}) \oplus \Lambda(R', 1, \tau, \begin{pmatrix} 0 & 1 \\ 0 & 0 \end{pmatrix})$, *which can be described as follows:*

$$\text{Let } \Gamma_0 = R \overset{\mathfrak{p}}{\longrightarrow} \begin{pmatrix} R & \mathfrak{p}^{1\times 2\tau} \\ R^{2\tau\times 1} & S^{\tau\times\tau} \end{pmatrix} \quad \text{and} \quad \Gamma_1 = \begin{pmatrix} \tilde{R} & \mathfrak{p}'^{1\times\tau} \\ R'^{\tau\times 1} & R'^{\tau\times\tau} \end{pmatrix}$$

where $S = \mathfrak{p}^{2\times 2} + R[\begin{pmatrix} 0 & 1 \\ 1 & 1 \end{pmatrix}]$ *as in (VII.1)(i) and* $\tilde{R} = \mathfrak{p}' + R1 \leq R'$. *Then there are epimorphisms* ν_0 *and* ν_1 *of* Γ_0 *and* Γ_1 *onto the same (finite) ring such that* $\Lambda_1 \cong \{ (a,b) \in \Gamma_0 \oplus \Gamma_1 \mid \nu_0(a) = \nu_1(b) \}$. *The kernels of* ν_1 *and* ν_2 *are*

$$\mathfrak{p} \overset{\mathfrak{p}}{\longrightarrow} \begin{pmatrix} \mathfrak{p} & (\mathfrak{p}^2)^{1\times 2\tau} \\ \mathfrak{p}^{2\tau\times 1} & \mathfrak{a}^{\tau\times\tau} \end{pmatrix} \quad \text{and} \quad \begin{pmatrix} \mathfrak{p}\tilde{R} & (\mathfrak{p}'^2)^{1\times\tau} \\ \mathfrak{p}'^{\tau\times 1} & (\mathfrak{p}'^2)^{\tau\times\tau} \end{pmatrix}$$

where $\mathfrak{p} \overset{\mathfrak{p}}{\text{———}} \mathfrak{p} := \mathfrak{p} \cdot (R \overset{\mathfrak{p}}{\text{———}} R)$ *and* $\mathfrak{a} = \mathfrak{p} \cdot \left\langle \begin{pmatrix} 1 & 1 \\ 0 & 1 \end{pmatrix}, \begin{pmatrix} 1 & 0 \\ 1 & 1 \end{pmatrix}, \mathfrak{p}^{2\times 2} \right\rangle_R \lhd S$ *as in (VII.1)(i).*

(ii) The epimorphic image Λ_2 *of the principal block ideal* Λ *of* RG , $G = SL_2(q)$, *is isomorphic to the R-order*

$$\begin{pmatrix} \tilde{\Omega} & \mathfrak{P}^{1\times\tau} \\ \Omega^{\tau\times 1} & \Omega^{\tau\times\tau} \end{pmatrix} \overset{(\mathfrak{P}, \mathfrak{p}')}{\longrightarrow} R'^{\tau\times\tau} \qquad (\subseteq \Lambda(\Omega, 1, \tau, \begin{pmatrix} 0 & 1 \\ 0 & 0 \end{pmatrix}) \oplus \Lambda(R', \tau, (0)))$$

where Ω *is the maximal order in the unique central division algebra over* K *of index* 2 *with* $\mathfrak{P} = Jac(\Omega)$, *and* $\tilde{\Omega} = \mathfrak{P} + R1 \subseteq \Omega$.

Proof: The proof is analogous to the one of (VII.1) and can be omitted. Note, in the case of Λ_1 the amalgamating factor of a projective indecomposable Λ_1-lattice has composition factors, which are not all isomorphic, as one sees already from the amalgamation matrices in (VII.3). (This, however, does not affect the argument of translating (VII.3) in (VII.4).) In the case of the projective cover of S_1 this can nicely be observed by inducing up the relevant lattices from the Borel subgroup of $\bar{G}$.

q.e.d.

From (VII.4)(ii) or (i) a straightforward calculation yields the next result, cf. [Erd 77] for the ground field F algebraically closed.

(VII.5) Corollary. *Let X_1 and X_τ be the projective indecomposable b-modules with $X_i/Jac(X_i) \cong S_i$ $(i = 1,\tau)$, where $b \cong \Lambda_1/\mathfrak{p}\Lambda_1 \cong \Lambda_2/\mathfrak{p}\Lambda_2$ is the principal (2-)block of $F\bar{G}$, then the submodules of X_1 and X_τ can be read off from the following diagrams of the submodule lattices (note, $\dim_F S_\tau = 2\tau$):*

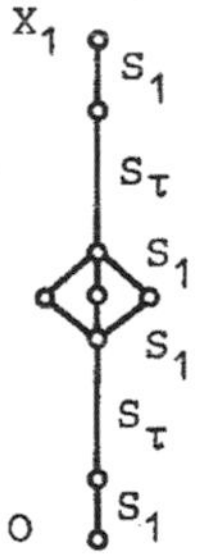

fig. 5

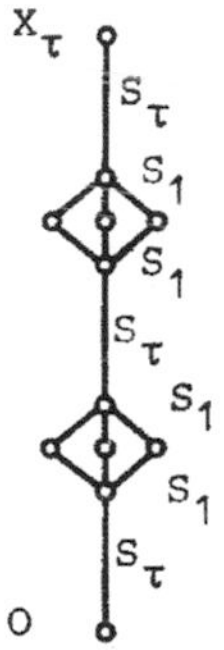

fig. 6

Case iii: $q \equiv 7 \bmod 8$

In this case F is a splitting field of $\Lambda/\mathfrak{p}\Lambda$.

The usual procedure (using (III.13)) shows that all Schur indices of the components of A are equal to 1 and that the decomposition numbers of Λ in the modified form as defined in (III.10) are given by:

	1	τ_1	τ_2
1	1	0	0
q	1	1	1
τ_1	0	1	0
τ_2	0	0	1
$(q-1)_2$	0	1	1
$(q-1)_3$	0	1	1
	⋮	⋮	⋮
$(q-1)_{n-1}$	0	1	1
$(\tau+1)_1$	1	1	0
$(\tau+1)_2$	1	0	1
$(q-1)_n$	0	1	1

The part of the matrix above the dotted line refers to Λ_1 , and the one below to Λ_2 resp. to the faithful Frobenius characters of G . The parameter $n \in \mathbb{N}$ is defined by $q \equiv 2^n-1 \pmod{2^{n+1}}$. The component of A corresponding to $(q-1)_i$ (cf. arguments based on (III.13) used in the earlier cases) is a matrix ring of degree $q-1$ over $K[\zeta_i + \zeta_i^{-1}]$ where ζ_i is a primitive 2^i-th root of unity. (Hence 2^{i-2} absolutely irreducible Frobenius characters belong to $(q-1)_i$, $2 \leq i \leq n$) .

(VII.6) Theorem. *Let* $q \equiv 2^n-1 \bmod 2^{n+1}$ *for some* $n \in \mathbb{Z}_{\geq 3}$, *R the ring of 2-adic integers, and* $\tau = \frac{q-1}{2}$. *The epimorphic image* Λ_2 *of the principal block ideal* Λ *of* RG , $G = SL_2(q)$, *is isomorphic to the R-order*

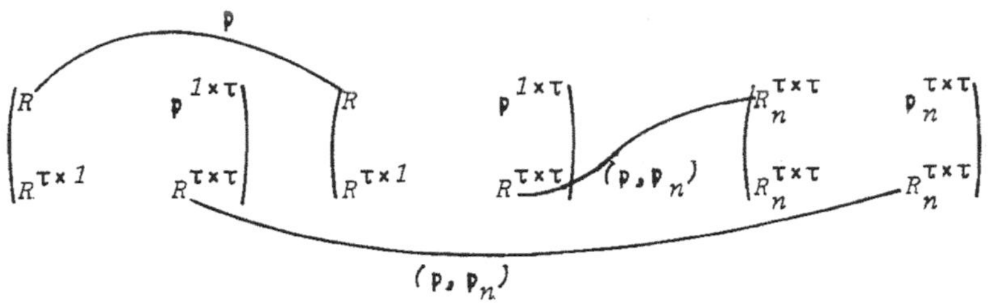

where $R_n = R[\zeta_n + \zeta_n^{-1}]$ *with* ζ_n *a primitive* 2^n*-th root of unity and* $\mathfrak{p}_n = Jac(R_n)$.

<u>Proof:</u> Clearly the epimorphic images of Λ_2 corresponding to $(\tau+1)_1$ and $(\tau+1)_2$ are graduated orders of the form $\Lambda(1,\tau,\begin{pmatrix}0 & a_i\\ 0 & 0\end{pmatrix})$ for suitable $a_1, a_2 \in \mathbb{N}$. In the case of $(q-1)_n$ one checks by inspecting the central characters of G that the center of Λ_2 is mapped onto the center of the epimorphic image of Λ_2 which is isomorphic to R_n (note R_n is the maximal R-order in its quotient field.) Hence (by (III.12)) one obtains a graduated order of the form $\Lambda(R_n,\tau,\tau,\begin{pmatrix}0 & b\\ 0 & 0\end{pmatrix})$ for some natural number b .
The amalgamation matrices of the projective Λ_2-covers $P_1^{(2)}$ and $P_{\tau_1}^{(2)}$ of S_1 and S_{τ_1} are

$\alpha(P_1^{(2)})$	1	τ_1	τ_2
$(\tau+1)_1$	1	$1-a_1$	0
$(\tau+1)_2$	1	0	$1-a_2$

and

$\alpha(P_{\tau_1}^{(2)})$	1	τ_1	τ_2
$(\tau+1)_1$	$1-a_1$	1	0
$(q-1)_n$	0	$\varkappa$	$\varkappa-b$

with $\varkappa = (n-1)\, 2^{n-2} - d_n$, where $\mathfrak{p}_n^{d_n}$ is the different of R_n with respect to R , (note 2^n is the highest power of 2 dividing $|\bar{G}|$, cf. (III.2) and (III.8)) .

By (IV.7) $a_1 = a_2 = \varkappa = b = 1$. Moreover the exponent matrices of the conducter of the graduated hull in Λ_2 are all three equal to $\begin{pmatrix}1 & 1\\ 0 & 1\end{pmatrix}$. This proves the theorem.

q.e.d.

Note, the proof of the theorem yields also the different of R_n with respect to R, namely $\mathfrak{p}_n^{d_n}$ with $d_n = (n-1)2^{n-2}-1$ as already obtained in (III.15)(iii). However, to conclude this for general n, one has to remark that an odd prime (power) q with $q \equiv 2^n-1 \bmod 2^n$ exists by Dirichlet's theorem on primes in arithmetic progressions. As in case (i) and (ii) one obtains a description of the projective indecomposable b-modules.

(VII.7) Corollary. *(cf. [Erd 77]): Let the b-modules X_1, X_{τ_1} and X_{τ_2} be the projective (b-)covers of the simple b-modules S_1, S_{τ_1} and S_{τ_2}, where $b(\cong \Lambda_1/\mathfrak{p}\Lambda_1 \cong \Lambda_2/\mathfrak{p}\Lambda_2)$ is the principal block of $F\bar{G}$. Then*

(a) $Jac(X_1) \;/\; Soc(X_1) \cong S_{\tau_1} \oplus S_{\tau_2}$

(b) $Jac(X_{\tau_i}) \;/\; Soc(X_{\tau_i}) \cong S_1 \oplus U_i$ for $i = 1,2$, where U_i is uniserial of length 2^{n-2} with composition factors S_{τ_1} and S_{τ_2} in alternating order, starting and ending with S_{τ_j}, $j \neq i$, $j = 1,2$.

(VII.8) Theorem. *Let $q \equiv 2^n-1 \pmod{2^{n+1}}$ for $n \in \mathbb{Z}_{\geq 3}$ as in (VII.6). The epimorphic images $\varepsilon_s\Lambda_1$ of the principal block ideal Λ_1 of $R\bar{G}$, $\bar{G} = PSL_2(q)$, which are R-orders in the simple components A_s of A^1, $s = 1, q, \tau_1, \tau_2, (q-1)_2, \ldots, (q-1)_{n-1}$ are given by:*

$$R\,,\ \Lambda(1,\tau,\tau,\begin{pmatrix} 0 & n & n \\ 0 & 0 & 1 \\ 0 & 1 & 0 \end{pmatrix})\,,\quad R^{\tau\times\tau}\,,\ R^{\tau\times\tau}\,,\ \text{and}$$

$\Gamma_i = \Lambda(R_i,\tau,\tau,\begin{pmatrix} 0 & 1 \\ 0 & 0 \end{pmatrix})$ for $i = 2,\ldots,n-1$, where $R_i = R[\zeta_i + \zeta_i^{-1}]$ with a primitive 2^i-th root of unity ζ_i, (note, $R_2 = R$).

Proof: Clearly $\varepsilon_s\Lambda_1$ for $s = 1, \tau_1, \tau_2$ are of the desired form (cf. decomposition numbers at the beginning of this section). The values of the central characters on the conjugacy classes of the elements of order $q+1$ of G show that the center of $\varepsilon_{(q-1)_i}\Lambda_1$ is R_i for $i=2,3,\dots,n-1$. Hence $\varepsilon_{(q-1)_i}\Lambda_1 \cong \Lambda(R_i,\tau,\tau,\begin{pmatrix}0 & a_i\\ 0 & 0\end{pmatrix})$ for suitable $a_i \in \mathbb{N}$, $i = 2,3,\dots,n-1$. Let $P_{\tau_1}^{(1)}$ be the projective Λ_1-cover of S_{τ_1}. By (VII.7)(b) $\varepsilon_i P_{\tau_1}^{(1)}/\mathfrak{p}\varepsilon_i P_{\tau_1}^{(1)}$ is uniserial for these i. Since R_i is ramified over R for $i \geq 3$, i.e. is a proper power of $\mathfrak{p}_i = \mathrm{Jac}(R_i)$, this uniseriality implies $a_3 = \dots = a_{n-1} = 1$. Because of the outer automorphism of G interchanging τ_1 and τ_2 (IV.1) implies that the exponent matrix of $\varepsilon_q\Lambda_1$ is $\begin{pmatrix}0 & b & b\\ 0 & 0 & c\\ 0 & c & 0\end{pmatrix}$. (IV.7)(i) implies $b=n$. To determine a_2 and c, one has to look at the amalgamation matrix of $P_{\tau_1}^{(1)}$, which by (III.8) is

$\alpha(P_{\tau_1}^{(1)})$	1	τ_1	τ_2
τ_1	0	n	0
q	0	n	$n-2c$
$(q-1)_2$	0	$n-1$	$n-1-a_1$
$(q-1)_3$	0	$2(n-3)+1$	$2(n-3)$
$\vdots$	$\vdots$	$\vdots$	$\vdots$
$(q-1)_i$	0	$2^{i-2}(n-i)+1$	$2^{i-2}(n-i)$
	$\vdots$		
$(q-1)_{n-1}$	0	$2^{n-3}+1$	2^{n-3}

Note, the different of R_i over R is $\mathfrak{p}_i^{d_i}$ with $d_i = (i-1)2^{i-2}-1$ by (III.15)(iii) or the comment after (VII.6). Since R_i is completely ramified over R, the ramification index e_i is 2^{i-2}, the

application of (IV.7)(iii) to the last column of $\alpha(P_{\tau_1}^{(1)})$ leaves two possibilities:

Either, one of the numbers $n-2c$ and $n-1-a_2$ is equal to $n-3$ and the other less or equal $n-3$, or both numbers $n-2c$ and $n-1-a_2$ are equal and bigger than $n-3$. In the last case, one has $c = a_2 = 1$ as desired. The first case one has $a_2 = 2$ which, however, leads to the following contradiction. The embedding of $\bar{G}$ into $PGL_2(q)$ yields an (outer) automorphism φ of $\bar{G}$ interchanging S_{τ_1} and S_{τ_2}. In case $a_2 = 2$, φ has to map the (by (II.23)) unique irreducible $R\bar{G}$-lattice L with $\varepsilon_{(q-1)_2} L = L$ and $L/Jac(L) \cong S_{\tau_1} \oplus S_{\tau_2}$ onto itself. This implies that the action of $\bar{G}$ on L can be extended to an action of $PGL_2(q)$ on L or (at least) on $R[\sqrt{-1}] \otimes_R L$. But the character table of $PGL_2(q)$, cf. [Ste 51], shows that the relevant absolutely irreducible Frobenius characters of $PGL_2(q)$ involve $\sqrt{2} = \zeta_3 + \zeta_3^{-1}$. which contradicts both.

q.e.d.

<u>*(VII.9) Theorem.*</u> *With the notation of (VII.8) the principal block ideal Λ_1 of $R\bar{G}$, $\bar{G} = PSL_2(q)$, $q \equiv 7 \pmod 8$, can be described as follows:*

Let $\Gamma_1 = R^{\tau\times\tau} \oplus R^{\tau\times\tau}$ and let $\nu_i : \Gamma_i \to T_i$ with $T_i = \Gamma_i/Jac(\Gamma_i)^{2^{i-1}-1}$ be the natural epimorphism for $i = 2,3,\dots,n-1$. Define R-orders $\Delta_1, \Delta_2, \dots, \Delta_{n-1}$ inductively: $\Delta_1 = \Gamma_1$; if Δ_i for one i, $1 \leq i < n-1$ is defined, then there exists an epimorphism $\varphi_i : \Delta_i \to T_{i+1}$ with kernel $\mathfrak{p}\Delta_i$. Define $\Delta_{i+1} = \{(a,b) \in \Delta_i \oplus \Gamma_{i+1} \mid \varphi_i(a) = \nu_{i+1}(b)\}$. Then (for suitable choice of φ_j) $\Delta_1, \dots, \Delta_{n-1}$ are epimorphic images of Λ_1, as well as the R-order

$$\Gamma_0 = R \overset{\mathfrak{p}^n}{\text{———}} \begin{pmatrix} R & (\mathfrak{p}^n)^{1\times\tau} & (\mathfrak{p}^n)^{1\times\tau} \\ R^{\tau\times 1} & R^{\tau\times\tau} & \mathfrak{p}^{\tau\times\tau} \\ R^{\tau\times 1} & \mathfrak{p}^{\tau\times\tau} & R^{\tau\times\tau} \end{pmatrix}$$

Let $\mathfrak{J}_0$ *be the ideal*

$$\mathfrak{J}_0 = R \overset{\mathfrak{p}}{\text{———}} \begin{pmatrix} R & (\mathfrak{p}^n)^{1\times\tau} & (\mathfrak{p}^n)^{1\times\tau} \\ R^{\tau\times 1} & (\mathfrak{p}^n)^{\tau\times\tau} & (\mathfrak{p}^{n-1})^{\tau\times\tau} \\ R^{\tau\times 1} & (\mathfrak{p}^{n-1})^{\tau\times\tau} & (\mathfrak{p}^n)^{\tau\times\tau} \end{pmatrix}$$

of Γ_0 *with natural epimorphism* $\nu : \Gamma_0 \to \Gamma_0/\mathfrak{J}_0$. *Then there exists an epimorphism* $\varphi : \Delta_{n-1} \to \Gamma_0/\mathfrak{J}_0$ *such that* $\Lambda_1 \cong \{ (a,b) \in \Gamma_0 \oplus \Delta_{n-1} \mid \nu(a) = \varphi(b) \}$.

Proof: By induction on i it will be proved first that Δ_i is an epimorphic image of Λ_1 for $i = 1,2,\ldots,n-1$. This follows from (VII.8) for $\Delta_1 = \Gamma_1$ and Δ_2 . Assume that Δ_i is an epimorphic image of Λ_1 for some i, $2 \leq i < n-1$. By (VII.8) also Γ_{i+1} is an epimorphic image of Λ_1 , and the two epimorphisms $\Lambda_1 \to \Delta_i$ and $\Lambda_1 \to \Gamma_{i+1}$ can be composed to an epimorphism of Λ_1 onto a subdirect sum $\tilde{\Delta}_{i+1}$ of Δ_i and Γ_{i+1} . The amalgamation of Δ_i and Γ_{i+1} must be of such a kind that the projective indecomposable $\tilde{\Delta}_{i+1}$-lattices $\tilde{P}_{i+1,l}$, $l = \tau_1$ or τ_2 , are uniserial mod $\mathfrak{p}$ because of (VII.7)(b). But $\tilde{P}_{i+1,l}$ is an amalgam of (the corresponding) projective indecomposable lattices $P_{i,l}$ and $\approx\!\!P_{i+1,l}$ of Δ_i and Γ_{i+1} resp., $l = \tau_1,\tau_2$, both viewed as non-projective Λ_1-lattices. (VII.7)(b) implies that $P_{i,l}/\mathfrak{p}P_{i,l}$ is isomorphic as Λ_1-module to the unique factor module of $\approx\!\!P_{i+1,l}$ of the same length, namely of length $1 + 2 + 2^2 + \ldots + 2^{i-1} = 2^i - 1$. Moreover, $P_{i,l}$ and $\approx\!\!P_{i+1,l}$ have no (Λ_1-) epimorphic images in common which are properly bigger than $P_{i,l}/\mathfrak{p}P_{i,l}$. Namely, the minimal factor modules of $P_{i,l}$ bigger than this one are no longer annihilated by $\mathfrak{p}$ and therefore cannot be isomorphic to $\approx\!\!P_{i+1,l}/\mathfrak{p}\approx\!\!P_{i+1,l}$, which is the unique factor module of $\approx\!\!P_{i+1,l}$ of length 2^i . Hence $\tilde{P}_{i+1,l}$ is an amalgam of $P_{i,l}$ and $\approx\!\!P_{i+1,l}$ with $P_{i,l}/\mathfrak{p}P_{i,l} \cong \approx\!\!P_{i+1,l}/\mathrm{Jac}(\Lambda_1)^{2^i-1}\approx\!\!P_{i+1,l}$ as amalgamating factor, and therefore $\tilde{\Delta}_{i+1} \cong \Delta_{i+1}$. In particular Δ_{n-1} is an epimorphic image. The rest follows by inspection of the amalgamation matrices of the projective indecomposable Λ_1-lattices.

q.e.d.

Case iv: $q \equiv 1 \bmod 8$

Also in this final case F is a splitting field for $\Lambda/\mathfrak{p}\Lambda$, (by the usual argument) the Schur indices of the A_s are all equal to 1 , and the decomposition numbers of Λ in the modified form as defined in (III.10) are given by:

	1	τ_1	τ_2
1	1	0	0
q	1	1	1
$(\tau+1)_1$	1	1	0
$(\tau+1)_2$	1	0	1
$(q+1)_2$	2	1	1
⋮	⋮	⋮	⋮
$(q+1)_{n-1}$	2	1	1
τ_1	0	1	0
τ_2	0	0	1
$(q+1)_n$	2	1	1

The part of the matrix above the dotted line refers to Λ_1 , the one below to Λ_2 . The parameter $n \in \mathbb{N}$ is defined by $q \equiv 2^n+1 \bmod 2^{n+1}$. The component of A corresponding to $(q+1)_i$ is a matrix ring over $K[\zeta_i+\zeta_i^{-1}]$, where ζ_i is a primitive 2^i-th root of unity. Since some of the decomposition numbers in the first column are bigger than 1, the results of Chapter IV are not directly applicable, and one has to go back to Chapter III.

(VII.10) Theorem. *Let $q \equiv 2^n + 1 \bmod 2^{n+1}$ for some $n \in \mathbf{Z}_{\geq 3}$, R the ring of 2-adic integers and $\tau = \frac{q-1}{2}$. The epimorphic image Λ_2 of the principal block ideal Λ of RG, $G = SL_2(q)$, is isomorphic to*

$$R^{\tau\times\tau} \qquad R^{\tau\times\tau} \qquad \begin{pmatrix} R_n & \mathfrak{p}_n^{1\times\tau} & \mathfrak{p}_n & \mathfrak{p}_n^{1\times\tau} \\ R_n^{\tau\times 1} & R_n^{\tau\times\tau} & \mathfrak{p}_n^{\tau\times 1} & \mathfrak{p}_n^{\tau\times\tau} \\ R_n & R_n & R_n & \mathfrak{p}_n^{1\times\tau} \\ R_n^{\tau\times 1} & R_n^{\tau\times\tau} & R_n^{\tau\times 1} & R_n^{\tau\times\tau} \end{pmatrix}$$

$(\mathfrak{p},\mathfrak{p}_n)$ $\qquad$ $\mathfrak{p}_n$ $\qquad$ $(\mathfrak{p},\mathfrak{p}_n)$

$(\subseteq \Lambda(\tau,(0)) \oplus \Lambda(\tau,(0)) \oplus \Lambda(R_n,1,\tau,1,\tau,H_4)$ with $H_4 = \begin{pmatrix} 0&1&1&1\\0&0&1&1\\0&0&0&1\\0&0&0&0 \end{pmatrix}$) where $R_n = R[\zeta_n + \zeta_n^{-1}]$, ζ_n a primitive 2^n-th root of unity, and $\mathfrak{p}_n$ the maximal ideal of R_n.

<u>Proof:</u> Because of the introductory remarks a graduated hull of Λ_2 is of the form $\Lambda(\tau,(0)) \oplus \Lambda(\tau,(0)) \oplus \Lambda(R_n,1,\tau,1,\tau,H)$ for a suitable matrix $H = (h_{ij}) \in \mathbf{Z}_{\geq 0}^{4\times 4}$. Applying (III.8) yields $\Lambda(\tau,(1)) \oplus \Lambda(\tau,(1)) \oplus \Lambda(1,\tau,1,\tau,J_4 - H^{tr})$ as conductor of the graduated hull in Λ_2, where the entries of $J_4 \in \mathbf{Z}^{4\times 4}$ are all equal to 1. (Note the different of R_n over R is $\mathfrak{p}_n^{d_n}$ with $d_n = (n-1)2^{n-2} - 1$, cf. comments after (VII.6) or (III.15)). Since $h_{ij} + h_{ji} > 0$ for $i \neq j$ one obtains $h_{ij} + h_{ji} = 1$, $1 \leq i < j \leq 4$, i.e. H is the exponent matrix of a hereditary order. Let L be an irreducible Λ_2-lattice belonging to the component of $(q+1)_n$. It follows that the sublattices of L are linearly ordered by inclusion. To prove that H can be chosen to be H_4, it suffices to prove that there are no sub-

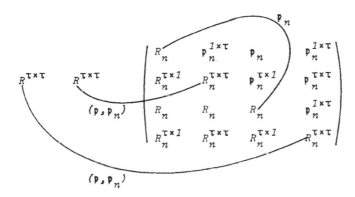

lattices L_1, L_2, L_3 of L with $L_1 \subset L_2 \subset L_3$ and $L_3/L_2 \cong L_2/L_1 \cong S_1$. But this is impossible, since the perfectness of G implies $\mathrm{Ext}^1_{FG}(S_1, S_1) = 0$. The rest follows from $\Lambda_2/\mathrm{Jac}(\Lambda_2) \cong F \oplus F^{\tau\times\tau} \oplus F^{\tau\times\tau}$.

q.e.d.

As in the previous cases one obtains a description of the projective indecomposable $F\bar{G}$-modules, in the principal block b.

(VII.11) Corollary (cf. [Erd 77]): *Let the b-modules X_1, X_{τ_1}, and X_{τ_2} be the projective (b-)covers of the simple b-modules S_1, S_{τ_1}, and S_{τ_2}, where $b (\cong \Lambda_1/\mathfrak{p}\Lambda_1 \cong \Lambda_2/\mathfrak{p}\Lambda_2)$ is the principal block of $F\bar{G}$. Then*

(i) $\mathrm{Jac}(X_1)/\mathrm{Soc}(X_1) \cong U_1 \oplus U_2$ where U_1 and U_2 are uniserial $F\bar{G}$-modules (of lengths 2^n-2) the composition factors of which come in the order $S_{\tau_1}, S_1, S_{\tau_2}, S_1, \ldots$ resp. $S_{\tau_2}, S_1, S_{\tau_1}, S_1, \ldots$.

(ii) X_{τ_1} and X_{τ_2} are uniserial, the composition factors come in the order $S_{\tau_1}, S_1, S_{\tau_2}, S_1, S_{\tau_1}, \ldots$ and $S_{\tau_2}, S_1, S_{\tau_1}, S_1, S_{\tau_2}, \ldots$ resp.

The version of Theorem (VII.5) for $q \equiv 1\ (8)$ is slightly more complicated.

(VII.12) Theorem. *Let $q \equiv 2^n + 1 \bmod 2^{n+1}$ for some $n \in \mathbb{Z}_{\geq 3}$, $\tau = \frac{q-1}{2}$, R the ring of 2-adic integers as in (VII.10), and let $R_i = R[\zeta_i + \zeta_i^{-1}]$ be the maximal order in $K[\zeta_i + \zeta_i^{-1}]$, ζ_i a primitive 2^i-th root of unity, with $\mathfrak{p}_i = \mathrm{Jac}(R_i)$ for $i = 2, \ldots, n-1$. The epimorphic images $\varepsilon_s\Lambda_1$ of the principal block ideal Λ_1 of $R\bar{G}$, $\bar{G} = PSL_2(q)$, which lie in the simple components A_s of A^1,*
$s = 1, q, (\tau+1)_1, (\tau+1)_2, (q+1)_2, \ldots, (q+1)_{n-1}$ are given by

$$R\ ,\ \Lambda\left(1,\tau,\tau,\begin{pmatrix}0&1&1\\0&0&1\\0&1&0\end{pmatrix}\right)\ ,\ \Lambda\left(1,\tau,\begin{pmatrix}0&1\\0&0\end{pmatrix}\right),\ \Lambda\left(1,\tau,\begin{pmatrix}0&1\\0&0\end{pmatrix}\right)\ ,\quad \text{and}$$

$$\Gamma_i = \begin{pmatrix} R_i & \mathfrak{p}_i^{1\times\tau} & \mathfrak{p}_i & \mathfrak{p}_i^{1\times\tau} \\ R_i^{\tau\times 1} & R_i^{\tau\times\tau} & \mathfrak{p}_i^{\tau\times 1} & \mathfrak{p}_i^{\tau\times\tau} \\ R_i & R_i^{1\times\tau} & R_i & \mathfrak{p}_i^{1\times\tau} \\ R_i^{\tau\times 1} & R_i^{\tau\times\tau} & R_i^{\tau\times 1} & R_i^{\tau\times\tau} \end{pmatrix} \subseteq \Lambda(R_i,1,\tau,1,\tau,H_4) \quad \text{with}\quad H_4 = \begin{pmatrix}0&1&1&1\\0&0&1&1\\0&0&0&1\\0&0&0&0\end{pmatrix}\ ,$$

$2 \leq i < n$ (note, $R_2 = R$) .

Proof: The values of the irreducible Frobenius characters or more precisely the corresponding central characters on the classes of order $q-1$ show that the center of $\varepsilon_s\Lambda$ is the maximal order in the center of $\varepsilon_j A$ for $s = (q+1)_2,\dots,(q+1)_{n-1}$.

Let $P^{(1)}_{\tau_1}$ be the projective Λ_1-cover of S_{τ_1} and define an irreducible Λ_1-lattice L_i by $L_i = \varepsilon_{(q+1)_i} P^{(1)}_{\tau_1}$ for $i = 2,\dots,n-1$. By (VII.11)(ii) $L_i/\mathfrak{p}L_i$ is uniserial. But $L_i \cong_{\Lambda_1} \mathfrak{p}_i L_i$ and $\mathfrak{p}$ is a proper power of $\mathfrak{p}_i$ for $i \geq 3$. Hence L_i is uniserial for $3 \leq i < n$. This shows that $\varepsilon_{(q+1)_i}\Lambda$ has the hereditary order given in the theorem as graduated hull. The isomorphism type of $\varepsilon_{(q+1)_i}\Lambda_1/\mathrm{Jac}(\varepsilon_{(q+1)_i}\Lambda_1)$ shows that Γ_i contains $\varepsilon_{(q+1)_i}\Lambda_1$. Since R_i is the center of $\varepsilon_{(q-1)_i}\Lambda_1$, i.e. $\varepsilon_{(q-1)_i}\Lambda_1$ is an R_i-order, it is now straightforward to verify $\Gamma_i = \varepsilon_{(q+1)_i}\Lambda_1$ for $3 \leq i \leq n-1$. The remaining orders $\varepsilon_s\Lambda_1$, $s = 1,q,(\tau+1)_1,(\tau+1)_2,(q+1)_2$ have R as their center. The outer automorphism of G interchanging τ_1 and τ_2 shows that the exponent matrix of $\varepsilon_q\Lambda_1$ is of the form $\begin{pmatrix}0&a&a\\0&0&b\\0&b&0\end{pmatrix}$ for suitable $a,b \in \mathbb{N}$. The embedding of

$\bar{G}$ into the symmetric group of degree $q+1$ shows $a=1$ by (VI.1), since $4 \nmid q+1$. But $a=1$ also implies $b=1$ by (II.2)(*). Also because of the outer automorphism $\varepsilon_{(\tau+1)_1}\Lambda_1$ and $\varepsilon_{(\tau+1)_2}\Lambda_1$ have the same exponent matrix $\begin{pmatrix} 0 & c \\ 0 & 0 \end{pmatrix}$ for some $c \in \mathbb{N}$. To prove $c=1$ let $L = \bar{L}^G$ be the RG-lattice induced up from the one-dimensional lattice $\bar{L}$ of the Borel subgroup B of upper triangular matrices of G such that the kernel of the action of B on $\bar{L}$ is of index 2. Clearly L is a faithful $(\varepsilon_{(\tau+1)_1} + \varepsilon_{(\tau+1)_2})\Lambda_1$-lattice with $L_{(i)} = \varepsilon_{(\tau+1)_i}L$ irreducible, $i=1,2$. Since $L/\mathfrak{p}L$ is isomorphic to a familiar permutation module, one sees $L/\mathrm{Jac}(L) \cong S_1 \cong L_{(i)}/\mathrm{Jac}(L_{(i)})$ for $i=1,2$. Assume $c>1$. Then there are Λ_1-epimorphisms of $L_{(i)}$ onto $R/\mathfrak{p}^2$; hence also from L onto $R/\mathfrak{p}^2$. But the action of $\bar{G}$ on L is monomial. The matrix of the epimorphism is of the form $(s_1,\ldots,s_{p+1})$ with $s_i \in \{\pm 1\}$. From this one sees that the epimorphism of L onto $R/\mathfrak{p}^2$ can be lifted to an epimorphism of L onto R, which is a contradiction. Hence $c=1$.

It remains to prove $\varepsilon_{(q+1)_2}\Lambda = \Gamma_2$. To this end it suffices to show that L_2 is uniserial, since then the argument applied for the $\varepsilon_s\Lambda$, $s = (q+1)_3,\ldots,(q+1)_{n-1}$ yields the desired result. As proved above, $L_2/\mathfrak{p}L_2$ is uniserial. Next it will be shown that there are no sub-lattices L_2' and L_2'' of L_2 with $L_2'' \subseteq L_2'$ and $L_2'/L_2'' \cong S_{\tau_1} \oplus S_{\tau_2}$. By (II.23) this is equivalent to the statement that $\varepsilon_{(q+1)_2}\hat{\Lambda}_1$ has $\begin{pmatrix} 0 & 1 \\ 0 & 0 \end{pmatrix}$ as exponent matrix, i.e. that $\varepsilon_2\hat{\Lambda}_1$ is hereditary. Here $\hat{\Lambda}_1 = \hat{\varepsilon}\Lambda_1\hat{\varepsilon}$ where $\hat{\varepsilon}$ is an idempotent of Λ_1 such that $\hat{\varepsilon} + \mathrm{Jac}(\Lambda_1)$ is the sum of the two central idempotents of $\Lambda_1/\mathrm{Jac}(\Lambda_1)$ corresponding to S_{τ_1} and S_{τ_2}. Note, by (III.6) $\hat{\Lambda}_1$ is a selfdual order; moreover all decomposition numbers of $\hat{\Lambda}_1$ are zero and one. Let $\begin{pmatrix} 0 & a_1 \\ 0 & 0 \end{pmatrix}$ be the exponent matrix for $\varepsilon_2\hat{\Lambda}_1$. The amalgamation matrix of the projective $\hat{\Lambda}_2$-lattice $\hat{\varepsilon}P_{\tau_1}^{(1)}$ is given by the last two columns of the amalgamation matrix of $\tilde{P}_{\tau_1}^{(1)}$ in the proof of (VII.8) with $c = 1$.

(Again) by (IV.7)(iii) one concludes $n-2 = n-1-a_1$, i.e. $a_1 = 1$.

To prove that L_2 is uniserial, let L be a sublattice of L_2 with $\mathfrak{p}L_2 \underset{\neq}{<} L \underset{\neq}{<} L_2$. Assume, there exists a maximal sublattice M of L with $M \nsubseteq \mathfrak{p}L_2$. Then $L/M \cong \mathfrak{p}L_2/\mathfrak{p}\mathrm{Jac}(L_2) \cong S_{\tau_1}$. Since L_2 has no sublattices with factor module $S_{\tau_1} \oplus S_{\tau_2}$, one sees that L cannot be maximal or second maximal in L_2. (Note $L_2/\mathfrak{p}L_2$ is uniserial with composition factors S_{τ_1}, S_1, S_{τ_2}, S_1.) In particular the second maximal sublattice of L_2 is isomorphic to $\varepsilon_{(q+1)_2}P_{\tau_2}^{(1)}$. Applying the analogous argument of L_2 to this lattice (with the rôles of S_{τ_1} and S_{τ_2} interchanged) or noting that a certain outer automorphism of G maps L_2 onto this lattice, now yield that L_2 is uniserial.

q.e.d.

<u>(VII.13) Theorem.</u> *Assume the hypotheses and notation of (VII.12). Define*

$$\Gamma_1 = \begin{pmatrix} R & \mathfrak{p}^{1\times\tau} \\ R^{\tau\times 1} & R^{\tau\times\tau} \end{pmatrix} \overset{\mathfrak{p}}{\frown} \begin{pmatrix} R & \mathfrak{p}^{1\times\tau} \\ R^{\tau\times 1} & R^{\tau\times\tau} \end{pmatrix}$$

and let $\mathfrak{J}_i$ be the ideal

$$\mathfrak{J}_i = \begin{pmatrix} \mathfrak{p}_i & \mathfrak{p}_i^{1\times\tau} & \mathfrak{p}_i & (\mathfrak{p}_i^{2})^{1\times\tau} \\ \mathfrak{p}_i^{\tau\times 1} & \mathfrak{p}_i^{\tau\times\tau} & \mathfrak{p}_i^{\tau\times 1} & \mathfrak{p}_i^{\tau\times\tau} \\ R_i & \mathfrak{p}_i^{1\times\tau} & \mathfrak{p}_i & \mathfrak{p}_i \\ R_i^{\tau\times 1} & R_i^{\tau\times\tau} & \mathfrak{p}_i^{\tau\times 1} & \mathfrak{p}_i^{\tau\times\tau} \end{pmatrix} \quad (\text{arc } \overset{\mathfrak{p}_i}{\frown} \text{ linking } (1,1)\text{, } (3,1) \text{ and } (1,3))$$

of Γ_i *for* $i = 2,\dots,n-1$ *, where* $R_i \overset{\mathfrak{p}_i}{\text{———}} \mathfrak{p}_i$ *denotes the amalgam of the* R_i*-lattices* R_i *and* $\mathfrak{p}_i$ *with amalgamating factor* $R_i/\mathfrak{p}_i$ *. Let* $\nu_i : \Gamma_i \to T_i$ *be the natural epimorphism with* $T_i = \Gamma_i/\mathfrak{p}_i^{2^{i-2}-1}\mathfrak{J}_i$ *for* $i = 2,\dots,n-1$ *. Define R-orders* $\Delta_1, \Delta_2, \dots, \Delta_{n-1}$ *inductively:* $\Delta_1 = \Gamma_1$ *; if* Δ_i *for one* i *,* $1 \leq i < n-1$ *is defined, then there exists an epimorphism* $\varphi_i : \Delta_i \to T_{i+1}$ *with kernel* $\mathfrak{p}\Delta_i$ *. Define* $\Delta_{i+1} = \{ (a,b) \in \Delta_i \oplus \Gamma_{i+1} \mid \varphi_i(a) = \nu_{i+1}(b) \}$ *. Then (for a suitable choice of* φ_i *)* $\Delta_1, \dots, \Delta_{n-1}$ *are epimorphic images of* Λ_1 *, as well as the R-order*

$$\Gamma_0 = R \overset{\mathfrak{p}}{\text{———}} \begin{pmatrix} R & \mathfrak{p}^{1\times\tau} & \mathfrak{p}^{1\times\tau} \\ R^{\tau\times 1} & R^{\tau\times\tau} & \mathfrak{p}^{\tau\times\tau} \\ R^{\tau\times 1} & \mathfrak{p}^{\tau\times\tau} & R^{\tau\times\tau} \end{pmatrix}$$

Let $\nu : \Gamma_0 \to \Gamma_0/\mathfrak{J}_0$ *be the natural epimorphism where* $\mathfrak{J}_0$ *is the ideal*

$$\mathfrak{J}_0 = \mathfrak{p}^{n-1} \left(R \overset{\mathfrak{p}}{\text{———}} \begin{pmatrix} R & \mathfrak{p}^{1\times\tau} & \mathfrak{p}^{1\times\tau} \\ R^{\tau\times 1} & \mathfrak{p}^{\tau\times\tau} & R^{\tau\times\tau} \\ R^{\tau\times 1} & R^{\tau\times\tau} & \mathfrak{p}^{\tau\times\tau} \end{pmatrix} \right)$$

of Γ_0 *. Then there exists an epimorphism* $\varphi : \Delta_{n-1} \to \Gamma_0/\mathfrak{J}_0$ *such that* $\Lambda_1 \cong \{ (a,b) \in \Gamma_0 \oplus \Delta_{n-1} \mid \nu(a) = \varphi(b) \}$ *.*

<u>Proof:</u> By induction on i it will be proved first that Δ_i is an epimorphic image of Λ_1 for $i = 1,2,\dots,n-1$. By (VII.12) this is easy to see for $\Delta_1 = \Gamma_1$. Assume that Δ_i is an epimorphic image of Λ_1 for some i , $1 \leq i < n-1$. By (VII.12) also Γ_{i+1} is an epimorphic image of Λ_1 , and the two epimorphisms $\Lambda_1 \to \Delta_i$ and $\Lambda_1 \to \Gamma_{i+1}$ define a homomorphism $\Lambda_1 \to \Delta_i \oplus \Gamma_{i+1}$, the image of which will be denoted by $\tilde{\Delta}_{i+1}$. The amalgamation of Δ_i and Γ_{i+1} to $\tilde{\Delta}_{i+1}$ must be such that the projective indecomposable $\tilde{\Delta}_{i+1}$-lattices

$\tilde{P}_{i+1,l}$, $l = \tau_1$ or τ_2 (covering S_l), are uniserial mod $\mathfrak{p}$ because of (VII.11)(ii). But $\tilde{P}_{i+1,l}$ is an amalgam of the corresponding projective indecomposable lattices $P_{i,l}$ and $\approx{P}_{i+1,l}$ of Δ_i and Γ_{i+1} resp., $l = \tau_1,\tau_2$ (both viewed as non-projective Λ_1-lattices). (VII.11)(ii) implies that $P_{i,l}/\mathfrak{p}P_{i,l}$ is isomorphic as Λ_1-module to the unique factor module of $\approx{P}_{i+1,l}$ of the same length, which is $2+4(1+2+\ldots+2^{i-1}) = 2^{i+1}-2$. Moreover, $P_{i,l}$ and $\approx{P}_{i+1,l}$ have no Λ_1-modules as epimorphic images properly bigger than $P_{i,l}/\mathfrak{p}P_{i,l}$ in common by the same annihilator argument as in the proof of (VII.9). Hence $\tilde{P}_{i+1,l}$ is an amalgam of $P_{i,l}$ and $\approx{P}_{i+1,l}$ with amalgamating factor $P_{i,l}/\mathfrak{p}P_{i,l} \cong \approx{P}_{i+1,l}/\mathrm{Jac}(\Lambda_1)^{2^{i+1}-2}\approx{P}_{i+1,l}$ for $l = \tau_1,\tau_2$.

Unlike to the proof of (VII.9) the projective $\tilde{\Delta}_{i+1}$-cover $\tilde{P}_{i+1,1}$ of S_1 has to be investigated, which is also an amalgam of the projective lattices $P_{i,1}$ and $\approx{P}_{i+1,1}$ of Δ_i and Γ_{i+1} covering S_1 (both viewed as non-projective Λ_1-lattices). Note, $\approx{P}_{i+1,1}$ is not irreducible as the description of Γ_{i+1} in (VII.12) shows. (It might be helpful to view $\approx{P}_{i+1,1}$ as the two columns in the description of Γ_{i+1} connected by $\overset{\mathfrak{p}_i}{\overline{}}$.) The clue is that all three $R\bar{G}$-lattices $\tilde{P}_{i+1,1}$, $P_{i,1}$, and $\approx{P}_{i+1,1}$ are induced up from certain lattices of the subgroup $\bar{B}$ of $\bar{G}$ which is the image of the (Borel) subgroup of upper triangular matrices in G .

Note, $\bar{B}$ has an epimorphism onto the cyclic group $C_{2^{n-1}} = \langle x\rangle$ the kernel of which has an order not divisible by 2. Consider $Y_j = R[\zeta_j]$, (ζ_j a primitive 2^j-th root of unity) as irreducible $RC_{2^{n-1}}$-lattice, where x acts via multiplication by ζ_j , $j = 0,1,\ldots,n-1$. Moreover, let W_j be the unique epimorphic image of the regular $RC_{2^{n-1}}$-lattice satisfying $K\otimes_R W_j \cong K\otimes_R(Y_1\oplus Y_2\oplus\ldots\oplus Y_j)$ for $j = 1,2,\ldots,n-1$. A small modification of (III.14)(i) shows that W_{i+1} is an amalgam of W_i and Y_{i+1} with amalgamating factor $W_i/\mathfrak{p}W_i \cong Y_{i+1}/Y_{i+1}(1-\zeta_{i+1})^{2^{i}-1}$.

All $RC_{2^{n-1}}$-modules can be viewed as $R\bar{B}$-modules (by pulling back the operation). All statements just made about the $RC_{2^{n-1}}$-modules remain true for the corresponding $R\bar{B}$-modules which will from now on be denoted by the same symbols. Coming back to the original situation, one obtains: $\tilde{P}_{i+1,1}$, $P_{i,1}$, and $\tilde{\tilde{P}}_{i+1,1}$ are isomorphic (as Λ_1-lattices) to the induced modules $W_{i+1}{}^{\bar{G}}(:= R\bar{G} \otimes_{R\bar{B}} W_{i+1})$, $W_i{}^{\bar{G}}$, and $Y_{i+1}{}^{\bar{G}}$ respectively. Moreover, the whole diagrams describing the amalgamation can be "induced up", i.e. tensored with $R\bar{G}$. Hence $\tilde{P}_{i+1,1}$ is an amalgam of $P_{i,1}$ and $\tilde{\tilde{P}}_{i+1,1}$ with amalgamating factor $P_{i,1}/\mathfrak{p}P_{i,1} \cong$ $\cong \tilde{\tilde{P}}_{i+1,1}/\tilde{\tilde{P}}_{i+1,1}\, \rho_{i+1}{}^{2^{i}-1}$, where ρ_{i+1} is a Λ_1-endomorphism of $\tilde{\tilde{P}}_{i+1,1}$ (coming from the multiplication with $1-\zeta_{i+1}$) satisfying $\tilde{\tilde{P}}_{i+1,1}\, \rho_{i+1} = \mathfrak{p}_i \tilde{\tilde{P}}_{i+1,1}$.

Together with the description of the amalgamation of $\tilde{P}_{i+1,l}$, $l = \tau_1, \tau_2$, this now implies $\Delta_{i+1} = \tilde{\Delta}_{i+1}$ and finishes the induction.

The rest of the proof is straightforward by application of (III.8) and (III.7).

q.e.d.

VIII. Blocks with cyclic defect groups

The aim of this chapter is a description of the ring theoretical structure of a block ideal B of a group ring $\mathbb{Z}_pG$ over the p-adic integers $\mathbb{Z}_p$ such that the defect group D of B is cyclic of order p^a. Before this will be done, a block Λ of RG will be investigated where Λ is a direct summand of $R \otimes_{\mathbb{Z}_p} B$ and R is the maximal $\mathbb{Z}_p$-order in the minimal unramified extension field K of $\mathbb{Q}_p$ such that $F = R/\mathfrak{p}$ ($= \mathrm{Jac}(R)$) is the minimal splitting field of B/pB. For the investigation of the R-order Λ the basic notation of Chapter IV will be kept. A good deal of the theory of blocks with cyclic defect groups starting with J.S. THOMPSON's paper [Tho 67] will be assumed in this chapter. In particular DADE's results in [Dad 66] on the decomposition numbers and generalized decomposition numbers, as well as the results by KUPISCH [Kup 68] and JANUSZ [Jan 69] on the structure of the projective $\Lambda/\mathfrak{p}\Lambda$-modules (in the form stated by PEACOCK in [Pea 77]) will be used for the investigation of Λ. A generalization of BRAUER's Theorem 11 in [Bra 41] on the sublattices of irreducible Λ-lattices, where Λ has defect one, to the above case will be obtained. This was mentioned as an open problem at the end of [Dad 66] and will be answered here by describing the $\varepsilon_s\Lambda$, $1 \leq s \leq h$. Moreover a description of Λ in terms of the amalgamation of the $\varepsilon_s\Lambda$ will be given. For the Galois descent from Λ to B some results on the fields of character values given in [Fei 82], cf. also [Ben 76], will be used.

Let $\hat{K} = K[\zeta]$ for some primitive $|G|$-th root ζ of unity, $\hat{R}$ the maximal R-order in $\hat{K}$, and $\hat{\Lambda} = \hat{R} \otimes_R \Lambda$. Because of the above choice of R, $\hat{\Lambda}$ is indecomposable as a ring, and K is equal to $\mathbb{Q}_p$ extended by the values of the Brauer characters belonging to $\hat{\Lambda}$. Dade

[Dad 66] introduced the stabilizer E of a certain block of $RC_G(D)$ in $N_G(D)$ such that $E/C_G(D)$ has order e, $e \mid p-1$ and such that $\frac{p^a-1}{e}+e$ respectively e is the number of irreducible Frobenius respectively Brauer characters belonging to $\hat{\Lambda}$. In the sequel the number e, called the inertial index of $\hat{\Lambda}$, as well as the subgroup $\bar{E}$ of Aut(D) isomorphic to $E/C_G(D)$ will play a rôle.

(VIII.1) Lemma: A contains $h = a + e$ components A_s, $s = 1,\dots h$. (After a suitable permutation of the indices) the centers of A_s are isomorphic to K for $s = a+1,\dots,h$, and completely ramified extension fields Z_s of K of degree $\frac{p^s-p^{s-1}}{e}$ for $s = 1,\dots,a$. The center of $\varepsilon_s\Lambda$ is the maximal order R_s of Z_s for $s = 1,\dots,a$ and isomorphic to R for $s = a+1,\dots,h$. Moreover the different of R_s over R is $\mathfrak{p}_s^{d_s}$ with $d_s = s\,\frac{p^s-p^{s-1}}{e} - \frac{p^{s-1}-1}{e} - 1$ for $s = 1,\dots,a$ where $\mathfrak{p}_s$ is the maximal ideal of R_s. (Note, $Z_1 = K$ iff $e = p-1$.).

Proof: It will be shown that the center $Z(\Lambda)$ of Λ is projected onto the maximal order R_s of Z_s by multiplication with ε_s, i.e. $\varepsilon_s Z(\Lambda) = R_s$ $(= Z(\varepsilon_s\Lambda))$ for $s = 1,\dots,h$. Let χ_s be one (there might be several) absolutely irreducible Frobenius character of $\hat{\Lambda}$ with $\chi_s(\varepsilon_s) \neq 0$. Denote the associated central character of G by ω_s; i.e. $\omega_s(\bar{x}) = \frac{|x|\chi_s(g)}{\chi_s(1)}$ for all $g \in G$, where $x = g^G$ is the conjugacy class of g and $\bar{x} \in RG$ the class sum. Then $\varepsilon_s Z(\Lambda)$ is isomorphic to $\tilde{R}_s$ which is defined as the extension of R by the values $\omega_s(\bar{x})$ with x running through all conjugacy classes G.

There are three types of conjugacy classes x:

(i) the elements of x are of p'-order,

(ii) the p-parts of the elements of x are not conjugate to an element $\neq 1$ of the defect group D, and

(iii) the p-parts of the elements of x are conjugate to an element

$\neq 1$ of D.

One need not consider the classes of type (i) because $\omega_s(\bar{x}) \in R$ for these classes x by the choice of R. For the classes x of type (ii) one has $\omega_s(\bar{x}) = 0$ by Brauer's Second Main Theorem on blocks. The classes of type (iii) remain to be discussed. For these the character values $\chi_s(g)$ can be computed in terms of the generalized decomposition numbers, given by Dade in [Dad 66], Theorem 1, Part 3, and in terms of the Brauer characters of $C_G(D_i)$ where D_i is the subgroup of index p^i for some i, $0 \leq i < a$ such that the p-part of g is conjugate to a generator of D_i. If χ_s is a non-exceptional character, cf. [Dad 66], then the character values lie in R by [Dad 66], Lemma 8.11. Since there are e non-exceptional characters, cf. [Dad 66], Theorem 1, Part 1, the claim of (VIII.1) for the last e of the orders $\varepsilon_s\Lambda$ is proved, i.e. $R = \tilde{R}_s$ in these cases.

To prove $\tilde{R}_s = R_s$ in the other cases let $K' \subseteq \hat{K}$ be an unramified extension field of K which contains the values of the Brauer characters of the centralizers $C_G(D_i)$ for $i = 0,1,\ldots,a-1$. To prove $\tilde{R}_s = R_s$ it suffices to show $R'\tilde{R}_s = R'_s$ where R' is the maximal R-order in K' and where R'_s denotes the maximal R-order in the field K'_s defined as follows. Let ζ_s be a primitive p^s-th root of unity in $\hat{K}$. Then $K'[\zeta_s]$ is a Galois extension of K' with Galois group isomorphic to $\mathrm{Aut}(D_{a-s})$ (note, D_{a-s} is cyclic of order p^s). The subgroup $\bar{E}$ of $\mathrm{Aut}(D)$ (introduced before (VIII.1)) can be identified with a unique subgroup of this Galois group; the corresponding fixed subfield of $K'[\zeta_s]$ is denoted by K'_s. $(1 \leq s \leq a)$. First note that the maximal order of $K'[\zeta_s]$ is generated as R'-module by the powers of ζ_s. Hence the maximal order R'_s is generated as R'-module by the orbit sums of these powers under the action of $\bar{E}$. But up to sign these orbit sums are the generalized decomposition numbers of χ_s as given in [Dad 66] Thm. 1, Part 3 (after suitable renumbering the indices). Therefore one only has to show that the generalized decomposition numbers can be written as R'-linear combinations of the values of the central character ω_s on the classes x of type (iii). First they can be written as R'-linear combinations of the $\chi_s(g)$ with (s fixed and) g lying in a class of type (iii), since the tables of the Brauer characters of the centralizers $C_G(D_i)$ $(i = 0,1,\ldots,a-1)$ are R'-invertible R'-matrices by [Bra 56] (3E). Finally $\chi_s(g) = \frac{\chi_s(1)}{|x|}\,\omega_s(\bar{x})$ for $g \in x$, where $\frac{\chi_s(1)}{|x|} \in R$ if x is a conjugacy class of type (iii). (Note χ_s is of height zero and $D \subseteq C_G(x)$.) This finishes the proof of $R'\tilde{R}_s = R'_s$, and hence of $\tilde{R}_s = R_s$.

Clearly the degree of Z_s over K is bounded by the one of K'_s over K' , namely by $\frac{p^s-p^{s-1}}{e}$. These degrees add up to $\frac{p^a-1}{e}$, which is the number of exceptional characters in $\hat{\Lambda}$. Hence the degree of Z_s over K is equal to $\frac{p^s-p^{s-1}}{e}$ and all components of A are taken care of and one has a component corresponding to exceptional characters. The discriminants of the Z_s over K can be obtained from (III.15) (i) and (ii), since K is unramified over $\mathbb{Q}_p$.

q.e.d.

By [Dad 66] Theorem 1, Part 2 the decomposition numbers of $\hat{\Lambda}$ are all equal to zero and one. Hence one obtains that the (modified) decomposition numbers of Λ (cf. (III.10)) are also equal to zero and one, and the application of (III.13) yields that the Schur indices of the A_s are equal to one.

(VIII.2) Corollary. $\varepsilon_s\Lambda$ *is a graduated order in* A_s *for* $s = 1,\ldots,h$. *Moreover* A_s *is a full matrix algebra over* Z_s *for* $s = 1,\ldots,a$ *and over* K *for* $s = a+1,\ldots,a+e$.

For the determination of the exponent matrices a more detailed description of the decomposition numbers and a description of the projective indecomposable $\Lambda/\mathfrak{p}\Lambda$-lattices $P_i/\mathfrak{p}P_i$, $i=1,\dots,r=e$, are needed. According to [Dad 66] Thm 1, Part 2, one has (in the terminology introduced at the beginning of Chapter IV), $r_1=\dots=r_a$ and $|c_i|=2$ or $1+a$, depending on whether $\{1,\dots a\}\subseteq c_i$ or $\{1,\dots,a\}\not\subseteq c_i$, for $i=1,\dots,e$. One can associate a Brauer tree T with Λ . The vertices of T are indexed by $\{1,\dots,a\}$, $a+1,\dots,a+e$, and two vertices are connected by an edge iff the corresponding r_s's have a non-empty intersection (which then consists of one element). The vertices of T are in 1-1-correspondence with the indices $1,\dots,e$ representing the irreducible Brauer characters. Let T_o be the set of all vertices of T having an odd distance from the exceptional vertex $\{1,\dots a\}$ and let T_e be the set of the remaining vertices, (distance meaning the number of edges between two vertices.). Clearly T_o partitions the edges of T into $|T_o|$ sets, each of which consists of the edges adjacent to one vertex in T_o ; in other words $\{1,\dots,e\}=\dot{\bigcup}_{s\in T_o} r_s$.

Mutatis mutandis the same holds for T_e . According to [Pea 77] (cf. also [Kup 68], [Jan 69]) these two partitions come from the orbits r_s, $s\in T_o$ of ρ respectively r_s, $s\in T_e$ of δ on $\{1,\dots,e\}$, where ρ and δ are two permutations of $\{1,\dots,e\}$ connected with the submodule structure of $\overline{P}_i := P_i/\mathfrak{p}P_i$ $(i=1,\dots,e)$ as follows:

$\overline{P}_i$ has two uniserial submodules $X_i(\rho)$ and $X_i(\delta)$ with $X_i(\rho)+X_i(\delta)=\mathrm{Jac}(\overline{P}_i)$ and $X_i(\rho)\cap X_i(\delta)=\mathrm{Soc}(\overline{P}_i)$ such that the composition factors of $X_i(\rho)$ are (starting from the top) $S_{\rho(i)}$, $S_{\rho^2(i)}$, $\dots,S_i$ (composition factors S_j with $j\in r_s$ each with multiplicity one, where $s\in T_o$ with $i\in r_s$) and the composition factors of $X_i(\delta)$ are (also starting from the top) are $S_{\delta(i)}$, $S_{\delta^2(i)}$, $\dots$, S_i where the cycle of δ is repeated $\frac{p^a-1}{e}$ times in case $i \in r_1 = \ldots = r_a$ (exceptional vertex) and not repeated otherwise (composition factors S_j with $j \in r_t$ each with multiplicity $\frac{p^a-1}{e}$ if $r_t = r_1 = \ldots = r_a$ and with multiplicity one otherwise, where $t \in T_e$ with $i \in r_t$.).

(VIII.3) Theorem. Let Λ be a block ideal of RG with cyclic defect group of order p^a, such that the quotient field K of R is unramified over $\mathbb{Q}_p$ and that $F = R/\mathfrak{p}$ is a splitting field of $\Lambda/\mathfrak{p}\Lambda$. For $n \in \mathbb{N}$ let $H_n \in \mathbb{Z}^{n\times n}$ denote the matrix $H_n = \begin{pmatrix} 0 & 1 & \cdots & 1 \\ \vdots & \ddots & \ddots & \vdots \\ \vdots & & \ddots & 1 \\ 0 & \cdots & \cdots & 0 \end{pmatrix}$.

Then, in the notation introduced above,

$$\varepsilon_s\Lambda \cong \Lambda(R_s, n_i, n_{\delta(i)}, \ldots, n_{\delta^{|r_s|-1}(i)}, H_{|r_s|}) \quad (i \in r_s)$$

for $s = 1, \ldots, a$, *and*

$$\varepsilon_s\Lambda = \Lambda(n_i, n_{\sigma(i)}, \ldots, n_{\sigma^{|r_s|-1}(i)}, a\, H_{|r_s|})$$

for $s = a+1, \ldots, h = e+a$, *where* $i \in r_s$, $\sigma = \delta$ *if* $s \in T_e$ *and* $\sigma = \rho$ *if* $s \in T_o$.

Proof: One edge i of the Brauer graph connects exactly two vertices, exactly one of which, say s_i lies in T_o. Hence $\varepsilon_{s_i}P_i$ is irreducible; moreover $\varepsilon_{s_i}P_i/\mathfrak{p}\varepsilon_{s_i}P_i$ and $(1-\varepsilon_{s_i})P_i/\mathfrak{p}(1-\varepsilon_{s_i})P_i$ have S_i as only common modular constituent. This together with the description of the submodules of the $P_i/\mathfrak{p}P_i$ above yields that every projective indecomposable $\varepsilon_s\Lambda/\mathfrak{p}\varepsilon_s\Lambda$-module is uniserial. Hence, by (II.22) the exponent matrix of $\varepsilon_s\Lambda$ is of the form $a_sH_{|r_s|}$ (with the modular constituents arranged according to the cycles of δ resp. ρ) for some $a_s \in \mathbb{N}$. Since by the above remark about S_i one has $|c_i \cap c_j| \leq 1$ as long as the exceptional Frobenius characters are not involved in both c_i and c_j, i.e. $1 \notin c_i$, $1 \notin c_j$; (IV.7)(i) and (VIII.1) imply $a_{a+1} = \ldots = a_h = a$. Now let $1 \leq s \leq a$. If $R_s \neq R$, then $\mathfrak{p}R_s$ is a proper power of $\mathfrak{p}_s$ by (VIII.1), and $\varepsilon_sP_i/\mathfrak{p}\varepsilon_sP_i$ is uniserial (not only $\varepsilon_s P_i / \mathfrak{p}_s \varepsilon_s P_i$) for $i \in r_s$. This implies $a_s = 1$. The only s for which $R_s = R$ might occur is $s = 1$, namely in case $e = p-1$. It remains to prove $a_1 = 1$ in case $e = p-1$ and $|r_1| > 1$. Let $i,j \in r_1$, $i \neq j$. The non-zero entries in the j-th column of the amalgamation matrix $\alpha(P_i)$ of P_i are $a-a_1, pa-d_2-a, \ldots, p^s a-d_s-1, \ldots, p^a a-d_a-1$ where d_s is defined as in (VIII.1) with $e = p-1$; note, p^{s-1} is the degree of Z_s over K and Z_s is completely ramified over K. Assume $a > 1$. (In case $a = 1$ one gets immediately $a_1 = 1$ by (IV.7) (i); this case has already been treated in Chapter V.) Then (IV.7) (iii) implies $a-a_1 = a-1$, i.e. $a_1 = 1$.

q.e.d.

In the case $a = 1$ this theorem goes back to Brauer, cf. [Bra 41], Thm. 11, which states that for a block Λ of defect 1 (and R satisfying the above assumptions) all irreducible Λ-lattices are uniserial. For a version of this over the p-adics, cf. [Jac 81] and [Rog 80a]. The above result improves [Ple 80a] Satz (III.19), where the non-exceptional characters are treated under the assumption that K is a splitting field.

The next topic of this chapter concerns the amalgamation of the $\varepsilon_s\Lambda$. Since amalgamation matrices were already used in the proof of (VIII.3), their implication (via (IV.12)) for congruences of the central characters $\omega_{a+1}, \ldots, \omega_h$ might be mentioned first, cf. announcement in [Ple 80b].

<u>*(VIII.4) Corollary.*</u> *Omit the exceptional vertex and the edges adjacent to it from the Brauer tree T and call the resulting (union of) tree(s) truncated Brauer tree $\check{T}$. For $r,s \in \{e+1, \ldots, e+a = h\}$ the values of the central characters are congruent modulo $\mathfrak{p}^a = |D|R$, i.e.*

$$\frac{\chi_s(g)\,|g^G|}{\chi_s(1)} \equiv \frac{\chi_t(g)\,|g^G|}{\chi_t(1)} \pmod{\mathfrak{p}^a},$$

if s *and* t *belong to vertices lying in the same connected component of the truncated Brauer tree* $\check{T}$.

(VIII.5) Theorem. *Assume the hypothesis and the notation of (VIII.3).*

(i) *Let* $\nu_s : \varepsilon_s\Lambda \to X_s := \varepsilon_s\Lambda/Jac(\varepsilon_s\Lambda)^{x_s}$ *with* $x_s = |r_1|\frac{p^{s-1}-1}{e}$ *be the natural epimorphism for* $s = 2,\ldots,a$ *(note,* $\varepsilon_s\Lambda$ *is a hereditary order). Define R-orders* Γ_s *for* $s = 1,\ldots,a$ *inductively by* $\Gamma_1 = \varepsilon_1\Lambda$ *and if* Γ_{s-1} *is already defined for* $2 \leq s \leq a$ *, then there exists an epimorphism* $\varphi_{s-1} : \Gamma_{s-1} \to X_s$ *such that*
$\Gamma_s = \{(x,y) \in \Gamma_{s-1} \oplus \varepsilon_s\Lambda \mid \varphi_s(x) = \nu_s(y)\}$
is isomorphic to an epimorhic image of Λ *(namely* $(\varepsilon_1+\ldots+\varepsilon_s)\Lambda$*).*

(ii) *Let* $\Gamma_0 = (\varepsilon_{a+1}+\ldots+\varepsilon_{a+e})\Lambda$ *. Then* Γ_0 *splits into the direct sum of* τ *R-orders, where* τ *is the number of components of the truncated Brauer tree* $\check{T}$ *introduced in (VIII.4). The order* Γ_0 *can be visualised as follows: Whenever the vertices* s *and* t *of* T *are connected by an edge with the index* i $(1 \leq i \leq e,$ $i \notin r_1 = \ldots = r_a,$ $a < s,t \leq a+e = h)$ *then* $\tilde{\varepsilon}_i(\varepsilon_s\Lambda)\tilde{\varepsilon}_i$ *and* $\tilde{\varepsilon}_i(\varepsilon_t\Lambda)\tilde{\varepsilon}_i$ *can be identified with two copies of* $R^{n_i \times n_i}$*, and*
$\tilde{\varepsilon}_i\Gamma_0\tilde{\varepsilon}_i = \tilde{\varepsilon}_i\Lambda\tilde{\varepsilon}_i \cong (R \underset{\mathfrak{p}^a}{-} R)^{n_i \times n_i}$ *with*
$R \underset{\mathfrak{p}^a}{-} R = \{(x,y) \in R \oplus R \mid x \equiv y \bmod \mathfrak{p}^a\}$. *(*$n_i = dim_F S_i$*,* $\tilde{\varepsilon}_i$ *orthogonal idempotents of* Λ *such that* $\tilde{\varepsilon}_i + Jac(\Lambda)$ *are the central primitive idempotents of* $\Lambda/Jac(\Lambda)$ *as in Chapter IV). Moreover, these are the only amalgamations occurring, i.e.*

$$\tilde{\varepsilon}_i\Gamma_0\tilde{\varepsilon}_j = \tilde{\varepsilon}_i\Big(\bigoplus_{s=a+1}^{h} \varepsilon_s\Gamma_0\Big)\tilde{\varepsilon}_j \quad \text{for } i \neq j,\ 1 \leq i,j \leq e,\ i,j \notin r_1 .$$

(iii) *There are epimorphisms* μ *and* ν *of* Γ_a *and* Γ_0 *onto* $\bigoplus_{i \in r_1} (R/\mathfrak{p}^a)^{n_i \times n_i}$ *such that* $\Lambda \cong \{(x,y) \in \Gamma_a \oplus \Gamma_0 \mid \mu(x) = \nu(y)\}$.

Proof: (i) The proof is an induction on s. For $s=1$ the statement is trivial. Assume that it has been proved already for some $s-1$, $2 \leq s \leq a$. The projective indecomposable $(\varepsilon_1+\ldots+\varepsilon_s)\Lambda$-lattices $(\varepsilon_1+\ldots+\varepsilon_s)P_i$, $i \in r_1$ are uniserial mod $\mathfrak{p}$ by the results quoted before (VIII.3). Since $(\varepsilon_1+\ldots+\varepsilon_{s-1})P_i \,/\, \mathfrak{p}(\varepsilon_1+\ldots+\varepsilon_{s-1})P_i$ is uniserial of length $|r_i| \sum_{t=1}^{s-1} \frac{p^t-p^{t-1}}{e} = |r_i|\frac{p^{s-1}-1}{e}$ and $\varepsilon_s P_i/\mathfrak{p}\varepsilon_s P_i$ uniserial of length $|r_i|\frac{p^s-p^{s-1}}{e} > |r_i|\frac{p^{s-1}-1}{e}$ the amalgamating factor of $(\varepsilon_1+\ldots+\varepsilon_s)P_i$ as amalgam of $(\varepsilon_1+\ldots+\varepsilon_{s-1})P_i$ and $\varepsilon_s P_i$ is at least of length $|r_i|\frac{p^{s-1}-1}{e}$. But it cannot have a bigger length, because otherwise the annihilator in R of the amalgamating factor would be $\mathfrak{p}^2$ contradicting the uniseriality of $\varepsilon_s P_i$ and the strict inequality above. Statement (i) follows.

(ii) This follows immediately from the amalgamation matrices $\alpha(P_i)$ of the projective indecomposable Λ-lattices P_i with $i \notin r_1 = \ldots = r_a$, $1 \leq i \leq e$: They consist of two rows with zeroes in every column, except for the i-th column where the entries are equal to a. That $R \frac{\quad}{\mathfrak{p}^a} R$ has to be chosen as described above, is a straightforward verification (which by the way is tantamount to $\chi_s(1)+\chi_t(1) \in \mathfrak{p}^a$ because $R \frac{\quad}{\mathfrak{p}^a} R$ is selfdual by (III.6)).

(iii) This follows from (i), (ii) and the amalgamation matrices for $\alpha(P_i)$, $i \in r_1 = \ldots = r_a$.

q.e.d.

For the Galois descent from Λ to the block B of the group ring over the p-adic integers $\mathbb{Z}_p$ the following results, cf. [Fei 82] page

333, Lemma 13.1 and 13.2, on the values of the characters in the block $\hat{\Lambda}$ (introduced at the beginning of this chapter) will be used:

(i) Adjoining the character values of an (irreducible) nonexceptional Frobenius character or Brauer character to $\mathbb{Q}_p$ always yields the field K defined at the beginning of this section.

(ii) The maximal unramified subfield $\tilde{K}$ of the field $\mathbb{Q}_p(\chi)$ for an exceptional character χ in the block $\hat{\Lambda}$ does not depend on χ, where $\mathbb{Q}_p(\chi)$ denotes the extension of $\mathbb{Q}_p$ by the values of χ.

In case $a=1$ and $e=p-1$ these two statements can be interpreted correctly as follows: (i) holds for the Brauer characters and $p-1$ of the p irreducible Frobenius characters in $\hat{\Lambda}$, cf. [Fei 82]. By (VIII.1) also the last Frobenius character in $\hat{\Lambda}$ has a field of values contained in K. Call this field $\tilde{K}$, and view this last character as exceptional character. Then (i) and (ii) holds without restriction.

It is clear that $\tilde{K}$ is a subfield of K. Denote the maximal $\mathbb{Z}_p$-order in $\tilde{K}$ by $\tilde{R}$, and let $\tilde{\mathfrak{p}}$ be the maximal ideal of $\tilde{R}$. Define $d=\dim_{\mathbb{Q}_p} K$, $\tilde{d}=\dim_{\mathbb{Q}_p} \tilde{K}$, and $m=\dim_K \tilde{K} = d/\tilde{d}$.

<u>*(VIII.6) Lemma.*</u> *Let ε be a central primitive idempotent of $\mathbb{Q}_p B (:= \mathbb{Q}_p \otimes_{\mathbb{Z}_p} B)$ and let χ be an irreducible Frobenius character in $\hat{\Lambda}$ with $\chi(\varepsilon) \neq 0$. If χ is a nonexceptional character, then the center $Z(\varepsilon B)$ of εB is isomorphic to R. If χ is an exceptional character, then εB is a hereditary R-order.*

Proof: Use the notation $\Theta = \varepsilon B$ and $H = \mathbb{Q}_p \Theta$ $(:= \mathbb{Q}_p \otimes_{\mathbb{Z}_p} \Theta)$.

Assume first that χ is nonexceptional. Since $Z(H) = K$ one sees that

$K\otimes_{\mathbb{Q}_p} H$ is isomorphic to d copies of H viewed as K-algebras. Let $R \otimes_{\mathbb{Z}_p} \Theta$ decompose into the direct sum of l indecomposable R-orders Θ_i $(i = 1,\ldots,l)$. Then the Θ_i are Galois conjugates of each other and hence l divides d. The claim $Z(\Theta) = R$ is tantamount to $l = d$. Note Θ_i is an epimorphic image of the R-order Γ_0 described in (VIII.5) (ii). $K\Theta_i$ is isomorphic to the direct sum of d/l copies of $H \cong K^{n\times n}$ for some $n \in \mathbb{N}$. If $d \neq l$, then $2l = d$, because of the form of the decomposition matrix of Λ, and $n \equiv -n \mod |G|R$ since the order of the Sylow-p-subgroups of G divides the degrees of projective indecomposable RG-lattices, cf. (III.16). For odd p this leads immediately to a contradiction, for $p = 2$ one gets a contradiction between $e = 1$ and $\Theta_i/\mathrm{Jac}(\Theta_i) \cong d/l$ copies of $\Theta/\mathrm{Jac}(\Theta)$ (note property (i) quoted above).

Assume that χ is exceptional. Θ is hereditary iff $R\otimes_{\mathbb{Z}_p} \Theta$ is hereditary. The components Θ_i $(i = 1,\ldots,l)$ of $R\otimes_{\mathbb{Z}_p}\Theta$ can be viewed as epimorphic images of the R-order Γ_a defined in (VIII.5) (i). The claim therefore follows if one shows that $K\Theta_i$ is a simple component of $K\otimes_{\mathbb{Q}_p} H$. But this is clear since all components of $K\otimes_{\mathbb{Q}_p} H$ are isomorphic and any two components of $K\Gamma_a$, namely $A_1,\ldots,A_a$, are non-isomorphic.

q.e.d.

<u>(VIII.7) Corollary.</u> Let ε *and* χ *be as in (VIII.6).*

(i) If χ *is a nonexceptional character, then* εB *can be viewed as an R-order and is isomorphic to one of the orders* $\varepsilon_s\Lambda$ *with* $s \in \{a+1,\ldots,a+e\}$ *described in (VIII.3).*

(ii) If χ *is an exceptional character, then* εB *is isomorphic to*

$\Lambda(\Omega_s, \tilde{n}, H_x)$, *where* Ω_s *is the maximal* $\tilde{R}$*-order in a division algebra* D_s *of index* m $(= \dim_{\tilde{K}} K)$ *with center* $\tilde{K}(\chi)$, $x = \frac{|r_s|}{m}$,

$$H_x = \begin{pmatrix} 0 & 1 & \cdots & 1 \\ \vdots & \ddots & \ddots & \vdots \\ \vdots & & \ddots & 1 \\ 0 & \cdots & \cdots & 0 \end{pmatrix} \in \mathbf{Z}^{x \times x} , \quad \tilde{n} = (n_i, n_{\delta(i)}, \ldots, n_{\delta^{x-1}(i)}) \quad (i \in r_s) \text{ for}$$

some s , $1 \le s \le a$. *(Note,* $r_1 = \ldots = r_a$*).*

Proof: (i) Since $Z(\varepsilon B) \cong R$ by (VIII.6) one can view εB as R-order and hence $R \otimes_{\mathbf{Z}_p} (\varepsilon B)$ is isomorphic to the direct sum of d copies of εB . Since $R \otimes_{\mathbf{Z}_p} B$ is isomorphic to the direct sum of Galois conjugates of Λ , the claim follows.

(ii) εB is hereditary with center isomorphic to the maximal $\mathbf{Z}_p$-order in $\tilde{K}(\chi)$. Since $\tilde{K}$ is the maximal unramified subfield of $\tilde{K}(\chi)$ (= $\tilde{K}$ extended by the values of χ), which is the center of $H = \mathbf{Q}_p \otimes_{\mathbf{Z}_p} (\varepsilon B)$ one sees that $K \otimes_{\mathbf{Q}_p} H$ is isomorphic to the direct sum of d simple K-algebras. The isomorphism type of the $\tilde{d}$ components of $R \otimes_{\mathbf{Z}_p} (\varepsilon B)$ can be read off from the isomorphism type of $\varepsilon_s \Lambda$ $(1 \le s \le a)$ given in (VIII.3). A simple computation yields the desired description of εB .

q.e.d.

Because of (VIII.7) (ii) and (VIII.1) one can compare the numbers of exceptional characters belonging to $\hat{\Lambda}$ and $\widehat{R \otimes_{\mathbf{Z}_p} B}$, and concludes that $\widehat{R \otimes_{\mathbf{Z}_p} B}$ and hence also $R \otimes_{\mathbf{Z}_p} B$ split up into the direct sum of $\tilde{d}$ $(= \dim_{\mathbf{Q}_p} \tilde{K})$ (algebraically conjugate) blocks. (For this result and the determination of the Schur indices over $\mathbf{Q}_p$ of the characters in $\hat{\Lambda}$ following from (VIII.7) cf. also [Ben 76] and [Fei 82].) Since the Galois group $Gal(K/\mathbf{Q}_p)$ is cyclic of order d , one gets $Gal(K/\tilde{K})$

as stabilizer of any component of $R\otimes_{\mathbb{Z}_p} B$.

(VIII.8) Lemma. *B can be viewed as $\tilde{R}$-order.*

Proof: A similar argument as above shows that $\tilde{R}\otimes_{\mathbb{Z}_p} B$ splits into the direct sum of $\tilde{d}$ algebraically conjugate blocks $B_1,\ldots,B_{\tilde{d}}$. The R-order homomorphism which is a composition of the natural embedding of $B(\cong 1\otimes_{\mathbb{Z}_p} B)$ into $\tilde{R}\otimes_{\mathbb{Z}_p} B$ and the projection of $\tilde{R}\otimes_{\mathbb{Z}_p} B$ onto B_i (for some i , $1 \leq i \leq \tilde{d}$) is an isomorphism. But B_i is an $\tilde{R}$-order.

q.e.d.

Hence one can identify $R\otimes_{\tilde{R}}B$ with Λ . The process of passing from B to $\Lambda = R\otimes_{\tilde{R}}B$ is best described by means of the various actions of $Gal(K/\tilde{K})$ on the components of $A = K\Lambda$ respectively on the corresponding index set $\{1,\ldots,h = a+e\}$ and on the components of $\Lambda/Jac(\Lambda)$ respectively on the corresponding index set $\{1,\ldots,e\}$. Let γ be a generator of $Gal(K/\tilde{K})$. Let $\tilde{e} = \frac{e}{m} = e\cdot(\dim_{\tilde{K}}K)^{-1}$. Then $\tilde{e}\in\mathbb{N}$ and the indices can be chosen in such a way that

$$\gamma(s) = \begin{cases} s & s = 1,\ldots,a \\ s+\tilde{e} & s = a+1,\ldots,a+e \end{cases} \qquad (\text{read } s+\tilde{e} \text{ as } s+\tilde{e}-e \text{ if } s+\tilde{e} > h)$$

describes the operation of γ on $\{1,\ldots,h\}$ and $\gamma(i) = i+\tilde{e} \pmod e$ for $i\in\{1,\ldots,e\}$ describes the operation on the other index set. Hence $\{1,\ldots,a,a+1,\ldots,a+\tilde{e}\}$ resp. $\{1,\ldots,\tilde{e}\}$ can be chosen as index sets for the components of $\tilde{K}B$ resp. $B/Jac(B)$. It was already observed in [Fei 82] that $Gal(K/\tilde{K})$ operates as a graph automorphism on the Brauer tree T such that the exceptional vertex stays fixed and the orbits on the other vertices and the edges are all of length m . Therefore the connected components of the truncated Brauer tree T are permuted by $Gal(K/\tilde{K})$. By the same argument as in the proof of (VIII.8) one gets the following statement.

<u>(VIII.9) Remark.</u> *Let* η *be the sum of those central primitive idempotents of* $\tilde{K}B$ *(=* $\tilde{K}\otimes_{\tilde{R}} B$*) belonging to the nonexceptional characters of* $\tilde{\Lambda}$ *(indices* $a+1,\ldots,a+\tilde{e}$*), then* ηB *can be viewed as an R-order. In particular the R-order* Γ_0 *described in (VIII.5) (ii) (which can be identified with* $R\otimes_{\tilde{R}}(\eta B)$ *is isomorphic to the direct sum of* m *copies of* ηB *.*

Clearly one gets a new tree $\tilde{T}$ by identifying the orbits of $\mathrm{Gal}(K/\tilde{K})$ on the vertices and on the edges of T . (T can be viewed as an m-fold covering of $\tilde{T}$ ramified at the exceptional vertex, cf. [Fei 82].) This relation can be pursued further. The two permutations δ and ρ introduced before (VIII.3) describe a planar embedding of T (in the sense that ρ and δ indicate in which circular order the edges of T are arranged around a vertex.) The tree $\tilde{T}$ also has such a planar embedding. Namely it follows from (VIII.7) and (VIII.3) that $\gamma\,(\in \mathrm{Gal}(K/\tilde{K}))$ commutes with ρ and δ : The elements of the exceptional cycle of δ are permuted among each other; the other cycles of δ and of ρ are permuted in orbits of m cycles. Now it is obvious that ρ and δ induce permutations $\tilde{\rho}$ and $\tilde{\delta}$ of the edges of the folded Brauer tree $\tilde{T}$. Since the version of (VIII.3) and (VIII.5) (ii) for B was given above already the reformulation in terms of $\tilde{T}$, $\tilde{\rho}$, and $\tilde{\delta}$ is left to the reader. As for the B-version of (VIII.5) (i), note that the idempotents $\varepsilon_1,\ldots,\varepsilon_a$ of $A = K\Lambda$ mentioned there already lie in $\tilde{K}B$ (with the obvious identification of B with a subset of Λ).

<u>(VIII.10) Theorem.</u> *Let* $\tilde{\nu}_s : \varepsilon_s B \to \varepsilon_s B/\mathrm{Jac}(\varepsilon_s B)^{x_s} =: \tilde{X}_s$ *be the natural epimorphism for* $s = 2,\ldots,a$ *with the same* $x_s = |r_1| \cdot \frac{p^{s-1}-1}{e}$ *as in*

(VIII.5) (i). Define $\tilde{R}$*-orders* $\tilde{\Gamma}_s$ *for* $s = 1,\dots,a$ *inductively by* $\tilde{\Gamma}_1 = \varepsilon_1 B$ *, and if* $\tilde{\Gamma}_{s-1}$ *is already defined for* $2 \leq s \leq a$ *, then there exists an epimorphism* $\tilde{\varphi}_{s-1} : \tilde{\Gamma}_{s-1} \to \tilde{X}_s$ *such that*

$\tilde{\Gamma}_s = \{ (a,b) \in \tilde{\Gamma}_{s-1} \oplus \varepsilon_s B \mid \tilde{\varphi}_s(a) = \tilde{\nu}_s(b)\}$ *is isomorphic to* $(\varepsilon_1 + \dots + \varepsilon_s) B$ *.* $(1 \leq s \leq a)$*.*

Proof: Use (VIII.5) (i), (VIII.7) (ii) and the fact that $\eta' B$ can be identified with $\tilde{R} \otimes_{\tilde{R}} (\eta' B)$ in $R \otimes_{\tilde{R}} (\eta' B) = \eta' \Lambda$, $\eta' = \varepsilon_1 + \dots + \varepsilon_s$, $1 \leq s \leq a$.

q.e.d.

The version of (VIII.5) (iii) for B is obvious now: Replace Γ_a by $\tilde{\Gamma}_a$, Γ_0 by ηB as described in (VIII.9) (ηB can certainly also be described by means of the truncation of the Brauer tree $\hat{T}$), and r_1 by $r_1 \cap \{1,\dots,\hat{e}\}$. Note, however, that the substitutes of the epimorphisms μ and ν there are only epimorphisms of $\tilde{R}$-algebras this time. It is clearly possible to recover a description of the projective indecomposables of B/pB analogous to the one of $\Lambda/p\Lambda$ quoted above and thereby one might also obtain some of the results in [Mic 75] and [Mic 76], cf. also [Fei 82] for arbitrary fields in characteristic p by starting out from this point. Moreover for the special case of blocks of defect 1, slight improvements of the results in [Jac 81] and [Rog 80a] can be obtained from the above description of blocks of $\mathbb{Z}_p G$ with cyclic defect group.

REFERENCES

J.L. ALPERIN
[Alp 72] Minimal Resolutions, Proc. of the Gainesville conference on finite groups, 1972.

M. BENARD
[Ben 76] Schur indices and cyclic defect groups, Ann. of Math. 103 (1976), 283-304.

H. BENZ
[Ben 61] Über eine Bewertungstheorie der Algebren und ihre Bedeutung für die Arithmetik, Schriftenreihe der Inst. f. Math. 9, Berlin 1961

H. BENZ and H. ZASSENHAUS
[BeZ 83] Über verschränkte Produktordnungen, to appear J. of Number Theory.

G. BIRKHOFF
[Bir 67] Lattice Theory (AMS Colloq. Publ. 25), Amer. Math. Soc., Providence, Rhode Island, 1967.

R. BRAUER
[Bra 41] Investigations on Group Characters, Ann. of Math., 42, 4 (1941), 936-958.
[Bra 56] Zur Darstellungstheorie der Gruppen endlicher Ordnung, Math. Zeitschr. 63 (1956), 406-44
[Bra 66] Some applications of the theory of blocks of characters of finite groups III, J. of Algebra 3 (1966), 225-255.

S. BRENNER
[Bre 67] Endomorphism Algebras of Vector Spaces with Distinguished Sets of Subspaces, J. of Algebra 6, 1 (1967), 100-114.

C.W. CURTIS and I. REINER
[CuR 81] Methods of Representation theory, vol. 1, J. Wiley & Sons, New York, Chichester, Brisbane, Toronto, 1981.

E.C. DADE
[Dad 66] Blocks with cyclic defect groups, Ann. of Math. 84, 2 (1966), 20-48.

L. DORNHOFF
[Dor 72] Group representation theory, part A and B, Dekker, New York 1972.

K. ERDMANN
[Erd 77] Principal blocks of groups with dihedral Sylow 2-subgroup, Comm. in Algebra, 5 (7) (1977), 665-694.

H.K. FARAHAT
[Far 62] On the natural representation of the symmetric groups, Proc. Glasgow Math. Assoc. 5 (1962), 121-136.

W. FEIT
[Fei 82] The Representation Theory of Finite Groups, North-Holland 1982.

J.A. GREEN
[Gre 74] Vorlesungen über modulare Darstellungstheorie endlicher Gruppen, Vorlesungen aus dem Mathematischen Institut Giessen, Heft 2, 1974.

J.E. HUMPHREYS
[Hum 72] Introduction to Lie Algebras and Representation Theory, Springer Verlag, New York, Heidelberg, Berlin, 1972.

B. HUPPERT
[Hup 67] Endliche Gruppen, I, Springer Verlag Berlin, Heidelberg, New York, 1967.

B. HUPPERT and N. BLACKBURN
[HuB 82] Finite Groups II, Springer Verlag, Berlin, Heidelberg, New York, 1982.

H. JACOBINSKI
[Jac 66] On extensions of lattices, Michigan Math. J. 13 (1966), 471-475.
[Jac 81] Maximalordnungen und erbliche Ordnungen, Vorlesungen Fachbereich Math. Univ. Essen, Heft 6, 1981.
[Jac 81b] Beziehungen zwischen modularer und ganzzahliger Darstellungstheorie, Vortrag Essen 1981.

G.D. JAMES
[Jam 73] The Modular Characters of the Mathieu Groups, J. of Algebra 27, 57-111 (1973).

G. JAMES and A. KERBER
[JaK 81] The Representation Theory of the Symmetric Group, Addison-Wesley, Reading, Massachusetts 1981

G.J. JANUSZ
[Jan 69] Indecomposable modules for finite groups, Ann. of Math. 89 (1969), 209-241.

V.A. JATEGAONKAR
[Jat 74] Global dimension of tiled orders over a discrete valuation ring, Trans. Am. Math. Soc. 196 (1974), 313-330.

M. KLEMM
[Kle 75] Über die Reduktion von Permutationsmoduln, Math. Z. 143 (1975), 113-117.

M. KNESER
[Kne 54] Zur Theorie der Kristallgitter, Math. Annalen 127 (1954), 105-106.

H. KUPISCH
[Kup 68] Projektive Moduln endlicher Gruppen mit zyklischer p-Sylow Gruppe, J. Algebra 10 (1968), 1-7.

E. MACAOGAIN
[Mac 76] Decomposition matrices of symmetric and alternating groups, Trinity College Dublin Research Notes, TCD 1976-10.

J. MCKAY
[McK 79] The non-abelian simple groups G, $|G| < 10^6$ - Charactertables, Comm. in Algebra 7 (13) (1979), 1407-1445.

G.O. MICHLER
[Mic 75] Green Correspondence between Blocks with Cyclic Defect Group II, Springer Lecture Notes in Math. 488, (Representations of Algebras, Ottawa 1974) (1975), 210-235.
[Mic 76] Green Correspondence between Blocks with Cyclic Defect Group I, J. of Algebra 39 no. 1 (1976), 26-51.
[Mic 81] On blocks with multiplicity one, Springer Lecture Notes in Math. 903 (Representations of Algebras, Puebla 1980)(1981), 242-256.

C. NASTACESCU and F. van OYSTAEYEN
[NaO 82] Graded Ring Theory, North-Holland Publishing Comp., Amsterdam, New York, Oxford,1982.

R.A. PARKER
[Par 82] Nullity Methods for the Computer Calculation of Modular Characters, in preparation.

R.M. PEACOCK
[Pea 77] Ordinary character theory in a block with a cyclic defect group, J. Algebra 44 (1977), 203-220.

M.H. PEEL
[Pee 71] Hook representations of the symmetric group, Glasgow Math. J. 12, 2 (1971), 136-149.

W. PLESKEN
[Ple 77] On absolutely irreducible representations of orders, 241-262 in [Zas 77].
[Ple 78] On reducible and decomposalbe representations of orders, J. reine angew. Math. 297 (1978), 188-210.
[Ple 80a] Gruppenringe über lokalen Dedekindbereichen, Habilitationsschrift Aachen 1980.
[Ple 80b] Projective lattices over group orders as amalgamations of irreducible lattices, Springer Lecture Notes in Math. 832 (Representation Theory II, Ottawa 1979) (1980), 438-448.

I. REINER
[Rei 75] Maximal Orders, Academic Press, London, New York, San Francisco 1975.

K.W. ROGGENKAMP
[Rog 70] Lattices over Orders II, Springer Lecture Notes in Math. 142, 1970.
[Rog 79] Private communication (1979).
[Rog 80a] Representation theory of blocks of defect 1, Springer Lecture Notes in Math. 832 (Representation Theory II, Ottawa 1979) 1980, 521-544.
[Rog 80b] Integral representations and structure of finite group rings, Séminaire de Mathématiques Supérieures, Les Presses de L'Université de Montreal 1980.
[Rog 81] Structure of integral group rings, preprint (1981).

J.L. ROTMAN
[Rot 79] An Introduction to homological algebra, Academic Press 1979.

W. RUMP
[Rum 81a] Systems of lattices in vector spaces and their invariants, Comm. in Algebra 9 (9) (1981), 893-932.
[Rum 81b] On tame irreducible representations of orders, Comm. in Algebra 9 (9) (1981), 933-974.

I. SCHUR
[Sch 07] Über die Darstellung der endlichen Gruppen durch gebrochene lineare Substitutionen, Gesammelte Abhandlungen I, Berlin, Heidelberg, New York 1973, 198-250.

J.P. SERRE
[Ser 77] Linear Representations of Finite Groups, Springer Verlag, New York, Heidelberg, Berlin 1977.

[Ser 79] Local Fields, Springer Verlag, New York, Heidelberg, Berlin 1979.

R. STEINBERG
[Ste 51] The representation of GL(3,q), GL(4,q), PGL(3,q), and PGL(4,q), Can. J. Math. 3 (1951), 225-235.

R.B. TARSY
[Tar 71] Global dimension of triangular orders, Proceed. Americ. Math. Soc. 28, 2 (1971), 423-426.

J.G. THOMPSON
[Tho 67] Vertices and Sources, J. Algebra 6 (1967), 1-6.

L.C. WASHINGTON
[Was 82] Introduction to Cyclotomic Fields, Springer Verlag, New York, Heidelberg, Berlin 1982.

A. WIEDEMANN and K.W. ROGGENKAMP
[WiR 82] Path orders of global dimension two, to appear J. of Algebra (1982).

H. ZASSENHAUS
[Zas 54] Über eine Verallgemeinerung des Henselschen Lemmas, Archiv d. Math. V (1954), 317-325.
[Zas 69] A condition for compatibility of submodules of a module, J. Number Theory 1 (1969), 467-476.
[Zas 75] Graduated Orders, unpublished manuscript, (1975).
[Zas 77] (Ed.), Number Theory and Algebra, Collected Papers Dedicated to Henry B. Mann, Arnold E. Ross, and Olga Taussky-Todd, Academic Press, New York, San Francisco, London 1977.

A.G. ZAVADSKII and V.V. KIRICENKO
[ZaK 77] Semimaximal rings of finite type, Mat. Sbornik, Tom 103 (145) (1977) No. 3, Math. USSR Sbornik, vol. 32 (1977) No. 3, 273-391.

SUBJECT INDEX